SLIPPING THE SURLY BONDS

GREAT QUOTATIONS

on

FLIGHT

DAVE ENGLISH

McGraw-Hill

New York   San Francisco   Washington, D.C.   Auckland   Bogotá
Caracas   Lisbon   London   Madrid   Mexico City   Milan
Montreal   New Delhi   San Juan   Singapore   Sydney   Tokyo   Toronto

Library of Congress Cataloging-in-Publication Data

Slipping the surly bonds: Great quotations on flight/Dave English.
      p.   cm.
    Includes index.
    ISBN 0-07-022016-6
    1.   Aeronautics—Quotations. maxims, etc   I.   English, Dave.
TL509.S547   1998
629.13–dc21   1997
000.0—dc00   98-16099
                                                  CIP

McGraw-Hill
A Division of *The McGraw-Hill Companies*

456789   DOC/DOC   9032109

ISBN 0-07-022016-6

The sponsoring editor for this book was Shelley Carr, the editing supervisor was Penny Linskey, and the production supervisor was Tina Cameron. It was designed by Michael Mendelsohn and set in Granjon by MM Design 2000, Inc.

Printed and bound by R. R. Donnelley & Sons Company.

This book is printed on recycled, acid-free paper
containing a minimum of 50% recycled, de-inked fiber.

This book is dedicated to the memory of Jim Bradley.

There were only two reasons he taught people to fly:

he loved people, and he loved flying.

CONTENTS

INTRODUCTION

Len Morgan—World War II pilot, B-747 captain, and long-time columnist for *Flying* magazine—sent me an e-mail last year: "You should publish those quotes as a book." I was in shock. Len Morgan, a hero possessed of that rare ability to translate the thrill of flight onto paper, had written me. I'd never met him, but somehow he had seen my rough collection of aviation quotes on the Internet.

I was getting a lot of e-mail. Pilots sharing clichés, aviation fans sharing favorite passages from books, notes from widows and their children remembering Dad, jokes from Australia, long letters from folks at Boeing and NASA. The one-page list grew into something that no one person could have compiled alone, a compendium of firsthand accounts covering the complete history of flight. Since finding out that McGraw-Hill agreed with Len, I've spent a lot of time in university libraries researching quotations, searching historic accounts, checking sources, and verifying spelling. Unfortunately, not every quote has two independent sources, so of course, corrections (and additions) are welcomed. As this "century of flight" draws to a close, I think it's important to record in one place people's original thoughts on maybe our most impressive of achievements. The original spelling is used where possible, but since America and England were at their most linguistically independent when the airplane (or aeroplane) was developed, this practice can confuse those who insist that there is only one way to spell a word.

If you are looking for more background on the people who were there first, I strongly recommend *Who's Who in Aviation History,* by William H. Longyard (Airlife Publishing Ltd., England). If you are looking for a great read, hunt out books by Richard Bach, Ernest K. Gann, or Antoine de Saint-Exupéry. If you are a new pilot, *Stick and Rudder* by Wolfgang Langewiesche has stayed continuously in print since 1944 for good reason. And English majors interested in the impact of aviation might try *The Flying Machine and Modern Literature,* by Laurence Goldstein (Bloomington: Indiana University Press).

I hope you will agree with another aviation author, Rod Machado, who wrote me, "Quotes have always offered succinct, power-packed bits of wisdom which are easily digested by the reader. There's much to be learned as well as enjoyed." You can reach me via McGraw-Hill or simply do what Len Morgan did: e-mail me at english@skygod.com.

Dave English

"I wish I could write well enough to write about aircraft. Faulkner did it very well in Pylon but you cannot do something someone else has done though you might have done it if they hadn't."

—*Ernest Hemingway*

ACKNOWLEDGEMENTS

I would like to thank all the people that e-mailed me with quotes, corrections, and suggestions; unfortunately the list is too long to publish here. The wonderful photographs taken at Chicago O'Hare were all shot (and are copyrighted by) Conn McCarthy, fellow airline pilot. The other photos come from NASA, the U.S. Air Force, and the U.S. National Archives.

Permission was kindly granted for publication of certain copyrighted passages by the following companies: Atlantic Monthly Press, Business and Commercial Aviation, Scientific American, and Yale University Press.

Excerpts from *Wind, Sand and Stars,* copyright 1939 by Antoine de Saint-Exupéry and renewed 1967 by Lewis Galantiere, reprinted by permission of Harcourt Brace & Company.

Excerpt from *Reaching for the Skies,* copyright© Ivan Rendall 1988, published by BBC Books, granted by Sheil Land Associates Ltd., London.

Excerpt from *Stick & Rudder: An Explanation of the Art of Flying,* by Wolfgang Langewiesche copyright 1944 renewed 1972 reproduced by permission of The McGraw Hill Companies.

Excerpt from *The Sled Driver,* by Brian Shul, reprinted by permission by MACH 1, Inc. Copyright 1996.

High Flight

Oh, I have slipped the surly bonds of earth
　　And danced the skies on laughter-silvered wings;
Sunward I've climbed, and joined the tumbling mirth
　　Of sun-split clouds—and done a hundred things
You have not dreamed of—wheeled and soared and swung
　　High in the sunlit silence. Hov'ring there,
I've chased the shouting wind along, and flung
　　My eager craft through footless halls of air.
Up, up the long, delirious burning blue
　　I've topped the windswept heights with easy grace
Where never lark, or even eagle flew.
　　And, while with silent, lifting mind I've trod
The high untrespassed sanctity of space,
　　Put out my hand, and touched the face of God.

—*Pilot Officer John Gillespie Magee, Jr., RCAF, September 3, 1941 (1922-1941). An American citizen born in Shanghai of missionary parents and educated at Britain's famed Rugby School, John Magee, Jr., entered the United States in 1939. He had won a scholarship to Yale University, but felt he must aid the cause of freedom, and so instead enlisted in the Royal Canadian Air Force in September 1940. He went to England to fly Spitfires. It was during a test flight at 30,000 feet that he was inspired to write* High Flight. *He was killed during a dogfight on December 11, 1941, at age 19.*

Magic

and

Wonder of Flight

When once you have tasted flight, you will forever walk the earth with your eyes turned skyward, for there you have been, and there you will always long to return.

—*Leonardo da Vinci.*

The most beautiful dream that has haunted the heart of man since Icarus is today reality.

—*Louis Blériot.*

The natural function of the wing is to soar upwards and carry that which is heavy up to the place where dwells the race of gods. More than any other thing that pertains to the body it partakes of the nature of the divine.

—*Plato,* Phaedrus.

Instead of our drab slogging forth and back to the fishing boats, there's a reason to life! We can lift ourselves out of ignorance, we can find ourselves as creatures of excellence and intelligence and skill. We can be free! We can learn to fly!

—*Richard Bach,* Jonathan Livingston Seagull.

Most gulls don't bother to learn more than the simplest facts of flight—how to get from shore to food and back again. For most gulls, it is not flying that matters, but eating. For this gull, though, it was not eating that mattered, but flight. More than anything else, Jonathan Livingston Seagull loved to fly.

—*Richard Bach,* Jonathan Livingston Seagull.

Pilots take no special joy in walking. Pilots like flying.

—*Neil Armstrong.*

It's wonderful to climb the liquid mountains of the sky, behind me and before me is God and I have no fears.

—*Helen Keller, at age 74, on flight around the world, news reports of February 5, 1955.*

Lovers of air travel find it exhilarating to hang poised between the illusion of immortality and the fact of death.

—*Alexander Chase,* Perspectives, *1966.*

My airplane is quiet, and for a moment still an alien, still a stranger to the ground, I am home.

—*Richard Bach,* Stranger to the Ground.

The airplane is just a bunch of sticks and wires and cloth, a tool for learning about the sky and about what kind of person I am, when I fly. An airplane stands for freedom, for joy, for the power to understand, and to demonstrate that understanding. Those things aren't destructible.

—*Richard Bach,* Nothing by Chance.

It is as though we have grown wings, which thanks to Providence, we have learnt to control.

—*Louis Bleriot,* Atlantic Monoplanes of Tomorrow.

Science, freedom, beauty, adventure: what more could you ask of life? Aviation combined all the elements I loved. There was science in each curve of an airfoil, in each angle between strut and wire, in the gap of a spark plug or the color of the exhaust flame. There was freedom in the unlimited horizon, on the open fields where one landed. A pilot was surrounded by beauty of earth and sky. He brushed treetops with the birds, leapt valleys and rivers, explored the cloud canyons he had gazed at as a child. Adventure lay in each puff of wind.

I began to feel that I lived on a higher plane than the skeptics of the ground; one that was richer because of its very association with the element of danger they dreaded, because it was freer of the earth to which they were bound. In flying, I tasted a wine of the gods of which they could know nothing. Who valued life more highly, the aviators who spent it on the art they loved, or these misers who doled it out like pennies through their antlike days? I decided that if I could fly for ten years before I was killed in a crash, it would be a worthwhile trade for an ordinary life time.

—*Charles A. Lindbergh,* The Spirit of St. Louis.

To put your life in danger from time to time . . . breeds a saneness in dealing with day-to-day trivialities.

—*Nevil Shute,* Slide Rule.

I fly because it releases my mind from the tyranny of petty things . . .

—*Antoine de Saint-Exupéry.*

If the heavens be penetrable, and no lets, it were not amiss to make wings and fly up, and some new-fangled wits should some time or other find out.

—*Robert Burton*, The Anatomy of Melancholy, *1621.*

What can you conceive more silly and extravagant than to suppose a man racking his brains, and studying night and day how to fly?

—*William Law*, A Serious Call to a Devout and Holy Life XI, *1728.*

How posterity will laugh at us, one way or other! If half a dozen break their necks, and balloonism is exploded, we shall be called fools for having imagined it could be brought to use: if it should be turned to account, we shall be ridiculed for having doubted.

—*Horace Walpole, letter to Horace Mann, June 24, 1785.*

Flying. Whatever any other organism has been able to do man should surely be able to do also, though he may go a different way about it.

—*Samuel Butler.*

The airplane has unveiled for us the true face of the earth.

—*Antoine de Saint-Exupéry*, Wind, Sand, and Stars.

Real flight and dreams of flight go together. Both are part of the same movement. Not A before B, but all together.

—*Thomas Pynchon*, Gravity's Rainbow.

The modern airplane creates a new geographical dimension. A navigable ocean of air blankets the whole surface of the globe. There are no distant places any longer: the world is small and the world is one.

—*Wendell Willkie.*

We want the air to unite the peoples, and not to divide them.

—*Lord Swinton.*

Unlike the boundaries of the sea by the shorelines, the "ocean of air" laps at the border of every state, city, town and home throughout the world.

—*Welch Pogue.*

I've never known an industry that can get into people's blood the way aviation does.

—*Robert Six, founder of Continental Airlines.*

Maybe it's sex appeal, but there's something about an airplane that drives investors crazy.

—*Alfred Kahn, airline economist and the "father of deregulation."*

The desire to fly is an idea handed down to us by our ancestors who, in their grueling travels across trackless lands in prehistoric times, looked enviously on the birds soaring freely through space, at full speed, above all obstacles, on the infinite highway of the air.

—*Wilbur Wright.*

whhheeeEEEEEEEEEEEEEEE! The scream of jet engines rises to a crescendo on the runways of the world. Every second, somewhere or other, a plane touches down, with a puff of smoke from scorched tyre rubber, or rises in the air, leaving a smear of black fumes dissolving in its wake. From space, the earth might look to a fanciful eye like a huge carousel, with planes instead of horses spinning round its circumference, up and down, up and down. Whhheeeeeeeeeee!

—David Lodge.

Dad, I left my heart up there.

—Francis Gary Powers, CIA U-2 pilot shot down
over the Soviet Union, describing his first flight at age 14.

As soon as we left the ground I knew I myself had to fly!

—Amelia Earhart (1937), after her first flight in an airplane,
a 10 minute sightseeing trip over Los Angeles, 1920.

I wanted to go higher than Rockefeller Center, which was being erected across the street from Saks Fifth Avenue and was going to cut off my view of the sky. . . . Flying got into my soul instantly but the answer as to why must be found somewhere back in the mystic maze of my birth and childhood and the circumstances of my earlier life. Whatever I am is elemental and the beginnings of it all have their roots in Sawdust Road. I might have been born in a hovel, but I determined to travel with the wind and stars.

—Jacqueline Cochran, The Stars at Noon, *1954.*

I could have gone on flying through space forever.

Major Yuri Gagarin, quoted in The New York Times, *April 14, 1958.*

I've had a ball.

—General Charles "Chuck" Yeager, describing his 30-year Air Force career.

To invent an airplane is nothing. To build one is something. To fly is everything.

—Otto Lilienthal.

You will begin to touch heaven, Jonathan, in the moment you touch the perfect speed. And that isn't flying a thousand miles an hour, or a million, or flying at the speed of light. Because any number is a limit, and Perfect speed, my son, is being there.

—Richard Bach, Jonathan Livingston Seagull.

There is no excuse for an airplane unless it will fly fast!

—Roscoe Turner.

Professor Focke and his technicians standing below grew ever smaller as I continued to rise straight up, 50 metres, 75 metres, 100 metres. Then I gently began to throttle back and the speed of ascent dwindled till I was hovering motionless in midair. This was intoxicating! I thought of the lark, so light and small of wing, hovering over the summer fields. Now man had wrested from him his lovely secret.

—Hanna Reitch, German test pilot describing her first helicopter flight.

I take the paraglider to the mountain or I roll Daisy out of her hangar and I pick the prettiest part of the sky and I melt into the wing and then into the air, till I'm just soul on a sunbeam.

—*Richard Bach,* Running From Safety, *1994.* Daisy *is Richard's Cessna 337.*

High spirits they had: gravity they flouted.

—*Cecil Day Lewis.*

This is all about fun. You can grab ahold of an airplane here, and literally take your life in both hands. One for the throttle and one for the stick, and you can control your own destiny, free of most rules and regulations. It may not be better than sex, but it's definitely better than the second time. Adrenaline is a narcotic; it may be a naturally induced narcotic, but it is a narcotic. And once you get it movin' around in there, it's a rush like none other, and when this puppy gets movin . . .

—*Alan Preston, air race pilot.*

Flying without feathers is not easy; my wings have no feathers.

—*Titus Maccius Plautus,* Paenulus, *Act V, Scene 2, circa 220 B.C. Original: Sine pennis volare hau facilest: meae alea pennas non habent.*

He rode upon a cherub, and did fly: yea, he did fly upon the wings of the wind.

—*Psalms 18:10, circa 150 B.C.*

The reason birds can fly and we can't is simply that they have perfect faith, for to have faith is to have wings.

—*Sir James Matthew Barrie.*

No bird soars too high, if he soars with his own wings.

—*William Blake.*

Fly and you will catch the swallow.

—*James Howell,* Proverbs, *1659.*

Sometimes I feel a strange exhilaration up here which seems to come from something beyond the mere stimulus of flying. It is a feeling of belonging to the sky, of owning and being owned—if only for a moment—by the air I breathe. It is akin to the well known claim of the swallow: each bird staking out his personal bug-strewn slice of heaven, his inviolate property of the blue.

—*Guy Murchie,* Song of the Sky, *1954.*

Travelers are always discoverers, especially those who travel by air. There are no signposts in the air to show a man has passed that way before. There are no channels marked. The flier breaks each second into new uncharted seas.

—*Anne Morrow Lindbergh,* North to the Orient, *1935*

The man who flies an airplane . . . must believe in the unseen.

—*Richard Bach.*

. . . the fundamental magic of flying, a miracle that has nothing to do with any of its practical purposes—purposes of speed, accessibility, and convenience—and will not change as they change.

—*Anne Morrow Lindbergh,* North to the Orient, *1935*

Flying is within our grasp. We have naught to do but take it.

—*Charles F. Duryea,* Learning How to Fly,
Proceedings of the Third International Conference on Aeronautics, 1894.

It will free man from the remaining chains, the chains of gravity which still tie him to this planet. It will open to him the gates of heaven.

—*Wernher von Braun, on the importance of space travel, February 10, 1958.*

What is it that makes a man willing to sit up on top of an enormous Roman candle, such as a Redstone, Atlas, Titan or Saturn rocket, and wait for someone to light the fuse?

—*Tom Wolfe,* The Right Stuff, *1979.*

It was quite a day. I don't know what you can say about a day when you see four beautiful sunsets. . . . This is a little unusual, I think.

—*Astronaut John Glenn Jr., 1962.*

There is no flying without wings.

—*French proverb.*

No need to teach an eagle to fly.

—*Greek proverb.*

No bird ever flew nonstop from New York to Tokyo, or raced 15 miles high at triple the speed of sound. But birds do something else. They do not conquer the air; they romance it.

—*Peter Garrison.*

The desire to fly is an idea handed down to us by our ancestors who, in their grueling travels across trackless lands in prehistoric times, looked enviously on the birds soaring freely through space, at full speed, above all obstacles, on the infinite highway of the air.

—*Wilbur Wright.*

Aviation records don't fall until someone is willing to mortgage the present for the future.

—*Amelia Earhart.*

To most people, the sky is the limit. To those who love aviation, the sky is home.

—*Anon.*

Real flight and dreams of flight go together. Both are part of the same movement. Not A before B, but all together.

—*Thomas Pynchon,* Gravity's Rainbow.

It is as though we have grown wings, which, thanks to Providence, we have learnt to control.

—*Louis Blériot,* Atlantic Monoplanes of Tomorrow, *1927.*

A pilot's business is with the wind, with the stars, with night, with sand, with the sea. He strives to outwit the forces of nature. He stares with expectancy for the coming of dawn the way a gardener awaits the coming of spring. He looks forward to port as to a promised land and truth for him is what lives in the stars.

—*Antoine de Saint-Exupéry,* Wind, Sand and Stars.

Flying has torn apart the relationship of space and time: it uses our old clock but with new yardsticks.

—*Charles A. Lindbergh.*

Pilots are a rare kind of human. They leave the ordinary surface of the world, to purify their soul in the sky, and they come down to earth, only after receiving the communion of the infinite.

—*José Maria Velasco Ibarra, President of Ecuador.*

Why does one want to walk wings? Why force one's body from a plane just to make a parachute jump? Why should man want to fly at all? People often ask these questions. But what civilization was not founded on adventure, and how long could one exist without it? Some answer the attainment of knowledge. Some say wealth, or power, is sufficient cause. I believe the risks I take are justified by the sheer love of the life I lead.

—Charles A. Lindbergh.

I learned that danger is relative, and that inexperience can be a magnifying glass.

—Charles A. Lindbergh.

Accuracy means something to me. It's vital to my sense of values. I've learned not to trust people who are inaccurate. Every aviator knows that if mechanics are inaccurate, they get lost—sometimes killed. In my profession life itself depends on accuracy.

—Charles A. Lindbergh.

I learned to watch, to put my trust in other hands than mine. I learned to wander. I learned what every dreaming child needs to know—that no horizon is so far that you cannot get above it or beyond it. These I learned at once. But most things come harder.

—Beryl Markham, West With the Night.

If you are a woman, and are coming to the flying field seeking stimulation, excitement and flattery, you had better stay away until flying is a little bit safer. If you are thinking that flying will develop character; will teach you to be orderly, well-balanced; will give you an increasingly wider outlook; discipline you, and destroy vanity and pride; enable you to control yourself more and more under all conditions; to think less of yourself and your personal problems, and more of sublimity and everlasting peace that dwell serene in the heavens—if you seek these latter qualities, and think on them exclusively, why—FLY!

—Margery Brown, Flying *magazine, 1929.*

When I'm up in the air, it's like I'm closer to heaven; I can't explain the feeling.

—Jeffrey Gagliano.

To fly a kite is to hold God's hand.

—Daniel C. Hawkins.

They shall mount up on wings as eagles.

—Isaiah 40:31.

Whether outwardly or inwardly, whether in space or time, the farther we penetrate the unknown, the vaster and more marvelous it becomes.

—Charles A. Lindbergh, Autobiography of Values.

We who fly do so for the love of flying. We are alive in the air with this miracle that lies in our hands and beneath our feet.

—*Cecil Day Lewis.*

A small machine is ideal for short flights, joy riding the heavens, or sight seeing among the clouds; but there is something more majestic and stable about the big bombers which a pilot begins to love. An exquisite community grows up between machine and pilot; each, as it were, merges into the other. The machine is rudimentary and the pilot the intellectual force. The levers and controls are the nervous system of the machine, through which the will of the pilot may be expressed—and expressed to an infinitely fine degree. A flying-machine is something entirely apart from and above all other contrivances of man's ingenuity.

The aeroplane is the nearest thing to animate life that man has created. In the air a machine ceases indeed to be a mere piece of mechanism; it becomes animate and is capable not only of primary guidance and control, but actually of expressing a pilot's temperament.

—*Sir Ross Smith, KBE,* National Geographic Magazine, *March 1921.*

Flying alone! Nothing gives such a sense of mastery over time, over mechanism, mastery indeed over space, time, and life itself, as this.

—*Cecil Day Lewis.*

Flying has always been to me this wonderful metaphor. In order to fly you have to trust what you can't see. Up on the mountain ridges where very few people have been I have thought back to what every flyer knows. That there is this special world in which we dwell that is not marked with boundaries; that's not a map. We're not hedged about with walls and desks. So often in an office the very worst thing that can happen is you could drop your pencil. Out there's a reminder that there are a lot worse things that can happen, and a lot greater rewards.

—Richard Bach, television interview.

I am alive. Up here with the song of the engine and the air whispering on my face as the sunlight and shadows play upon the banking, wheeling wings, I am completely, vibrantly alive. With the stick in my right hand, the throttle in my left, and the rudder beneath my feet, I can savor that essence from which life is made.

—Stephen Coonts, FLY! A Colorado Sunrise, a Stearman, and a Vision.

Aeronautics confers beauty and grandeur, combining art and science for those who devote themselves to it. . . . The aeronaut, free in space, sailing in the infinite, loses himself in the immense undulations of nature. He climbs, he rises, he soars, he reigns, he hurtles the proud vault of the azure sky . . .

—Georges Besancon, founder of the first successful aviation journal, L'Aerophile, *February 1902.*

What freedom lies in flying, what Godlike power it gives to men . . . I lose all consciousness in this strong unmortal space crowded with beauty, pierced with danger.

—Charles A. Lindbergh.

Until now I have never really lived! Life on earth is a creeping, crawling business. It is in the air that one feels the glory of being a man and of conquering the elements. There is an exquisite smoothness of motion and the joy of gliding through space. It is wonderful!

—Gabriele D'Annunzio, 1909.

He knew that we gave constant lip service to the dictates of safety and howled like Christians condemned to the arena if any compromise were made of it. He knew we were seekers after ease, suspicious, egotistic, and stubborn to a fault. He also knew that none of us would have continued our careers unless we had always been, and still were, helpless before this opportunity to take a chance.

—Ernest K. Gann, Fate Is the Hunter.

It's the most exciting thing you have ever done with your pants on!

—Stephen Coonts, Flight of the Intruder.

To fly! To live as airmen live! Like them to ride the skyways from horizon to horizon, across rivers and forests! To free oneself from the petty disputes of everyday life, to be active, to feel the blood renewed in one's veins—ah! that is life. . . . Life is finer and simpler. My will is freer. I appreciate everything more, sunlight and shade, work and my friends. The sky is vast. I breathe deep gulps of the fine clear air of the heights. I feel myself to have achieved a higher state of physical strength and a clearer brain. I am living in the third dimension!

—*Henri Mignoet,* L'Aviation de L'Amateur; Le Sport de l'Air, *1934.*

Air racing may not be better than your wedding night, but it's better than the second night.

—*Mickey Rupp, air racer and former Indianapolis 500 driver.*

When we walk to the edge of all the light we have and take the step into the darkness of the unknown, we must believe that one of two things will happen. There will be something solid for us to stand on or we will be taught to fly.

—*Patrick Overton.*

Aeronautics was neither an industry nor a science. It was a miracle.

—*Igor I. Sikorsky.*

Oh, that I had wings like a dove, for then would I fly away, and be at rest.

—*Psalms 55:6.*

I'm a new man. I go home exhilarated.

—President George Bush, after skydiving from 12,500 ft at age 72, March 1997.

You can always tell when a man has lost his soul to flying. The poor bastard is hopelessly committed to stopping whatever he is doing long enough to look up and make sure the aircraft purring overhead continues on course and does not suddenly fall out of the sky. It is also his bound duty to watch every aircraft within view take off and land.

—Ernest K. Gann, Fate Is the Hunter.

It is appearances, characteristics and performance that make a man love an airplane, and they, told truly, are what put emotion into one. You love a lot of things if you live around them, but there isn't any woman and there isn't any horse, not any before nor any after, that is as lovely as a great airplane, and men who love them are faithful to them even though they leave them for others.

—Ernest Hemingway.

As a young boy dreaming of becoming an airman, if I had a choice between becoming chief of staff of the Air Force or becoming a fighter ace, I would have chosen to become a fighter ace.

—General Thomas White, USAF Chief of Staff, 1973.

[Flying fosters] fantasies of childhood, of omnipotence, rapid shifts of being, "miraculous" moments; it stirs our capacity for dreaming.

—*Joyce Carol Oates, "Coming Home,"* USAir *magazine, May 1995.*

The best way of travel, however, if you aren't in any hurry at all, if you don't care where you are going, if you don't like to use your legs, if you don't want to be annoyed at all by any choice of directions, is in a balloon. In a balloon, you can decide only when to start, and usually when to stop. The rest is left entirely to nature.

—*William Pene du Bois,* The Twenty-one Balloons.

Suddenly the wind ceased. The air seemed motionless around us. We were off, going at the speed of the air-current in which we now lived and moved. Indeed, for us there was no more wind; and this is the first great fact of spherical ballooning. Infinitely gentle is this unfelt motion forward and upward. The illusion is complete: it seems not to be the balloon that moves, but the earth that sinks down and away . . .

Villages and woods, meadows and chateaux, pass across the moving scene, out of which the whistling of locomotives throws sharp notes. These faint, piercing sounds, together with the yelping and barking of dogs, are the only noises that reach one through the depths of the upper air. The human voice cannot mount up into these boundless solitudes. Human beings look like ants along the white lines that are highways; and the rows of houses look like children's playthings.

—*Alberto Santos-Dumont,* My Air-Ships, *1904.*

There's something in a flying horse,
There's something in a huge balloon.

—*William Wordsworth,* Peter Bell, *Prologue, Stanza 1.*

I have known today a magnificent intoxication. I have learnt how it feels to be a bird. I have flown. Yes I have flown. I am still astonished at it, still deeply moved.

—Le Figaro, *with regards to a balloon ride, 1908.*

The cockpit was my office. It was a place where I experienced many emotions and learned many lessons. It was a place of work, but also a keeper of dreams. It was a place of deadly serious encounters, yet there I discovered much about life. I learned about joy and sorrow, pride and humility, fear and overcoming fear. I saw much from that office that most people would never see. At times it terrified me, yet I could always feel at home there. It was my place, at that time in space, and the jet was mine for those moments. Though it was a place where I could quickly die, the cockpit was a place where I truly lived.

—*Brian Shul,* The Sled Driver.

Whether we call it sacrifice, or poetry, or adventure, it is always the same voice that calls.

—*Antoine de Saint-Exupéry.*

An Airman's Grace

Lord of thunderhead and sky
Who place in man the will to fly
Who taught his hand speed, skill and grace
To soar beyond man's dwelling place
You shared with him the Eagle's view
The right to soar, as Eagles do
The right to call the clouds his home
And grateful, through your heavens roam
May all assembled here tonight
And all who love the thrill of flight
Recall with twofold gratitude
Your gift of Wings, Your gift of Food.

—*Father John MacGillivary, Royal Canadian Air Force.*

We have shaken the surly bonds of Earth and are reaching for the skies.

—*Anon.*

Man's mind and spirit grow with the space in which they are allowed to operate.

—*Kraft A. Ehricke, rocket pioneer.*

Again I felt that overpowering rush of excitement which I found almost everyone has experienced who has seen a man fly. It is an exhilaration, a thrill, an ecstasy. Just as children jump and clap their hands to see a kite mount, so, when the machine leaves the ground and with a soaring movement really flies upon its speeding wings, one feels impelled to shout, to rush after it, to do anything which will relieve the overcharged emotion.

—*Harry Harper, describing Louis Blériot's departure for Dover, in the* Daily Mail, *July 26, 1909.*

Flying is a lot like playing a musical instrument; you're doing so many things and thinking of so many other things, all at the same time. It becomes a spiritual experience. Something wonderful happens in the pit of your stomach.

—*Dusty McTavish.*

All my life, I've never been able to get enough airplanes. This will keep me flying every day.

—*Astronaut Robert "Hoot" Gibson, commander of four space shuttle missions, on taking a job as a Southwest Airlines B-737 first officer, 1996.*

For pilots sometimes see behind the curtain, behind the veil of gossamer velvet, and find the truth behind man, the force behind a universe.

—*Richard Bach,* Biplane.

Before take-off, a professional pilot is keen, anxious, but lest someone read his true feelings he is elaborately casual. The reason for this is that he is about to enter a new though familiar world. The process of entrance begins a short time before he leaves the ground and is completed the instant he is in the air. From that moment on, not only his body but his spirit and personality exist in a separate world known only to himself and his comrades.

As the years go by, he returns to this invisible world rather than to earth for peace and solace. There also he finds a profound enchantment, although he can seldom describe it. He can discuss it with others of his kind, and because they too know and feel its power they understand. But his attempts to communicate his feelings to his wife or other earthly confidants invariable end in failure. Flying is hypnotic and all pilots are willing victims to the spell. Their world is like a magic island in which the factors of life and death assume their proper values. Thinking becomes clear because there are no earthly foibles or embellishments to confuse it. Professional pilots are, of necessity, uncomplicated, simple men. Their thinking must remain straightforward, or they die—violently.

The men in this book are fictitious characters but their counterparts can be found in cockpits all over the world. Now they are flying a war. Tomorrow they will be flying a peace, for, regardless of the world's condition, flying is their life.

—*Ernest K. Gann, Foreword to* Island in the Sky, *1944.*

Those rotary engines . . . the Le Rhones, the Monos, and the Clergets! They made a sort of crackling hiss, and always the same smell of castor oil spraying backwards down the 0 in a fine mist over your leather helmet and your coat. They were delightful to fly, the controls so light, the engines so smooth running. Up among the sunlit cumulus under the blue sky I could loop and roll and spin my Camel with the pressure of two fingers on the stick besides the button which I used as little as possible. Looping, turn off the petrol by the big plug cock upon the panel just before the bottom of the dive, ease the stick gently back and over you go. The engine dies at the top of the loop; ease the stick fully back and turn the petrol on again so that the engine comes to life five or six seconds later.

She would climb at nearly a thousand feet a minute, my new Clerget Camel; she would do a hundred and ten miles an hour. She would be faster, I thought, than anything upon the Western Front. . . . A turn to the left in the bright sun, keeping the hedge in sight through the hole in the top plane. A turn to the right. Now, turn in, a little high, stick over and top rudder, the air squirting in upon you sideways round the windscreen. Straight out, over the hedge, and down onto the grass. Remember that the Clerget lands very fast, at over forty miles an hour, and with that great engine in the nose the tail was light. Watch it . . . Lovely.

—*Nevil Shute,* The Rainbow and the Rose.

Racing planes didn't necessarily require courage, but it did demand a certain amount of foolhardiness and a total disregard of one's skin. . . . I would be flying now, but there's precious little demand for an elderly lady air racer.

—Mary Haizlip, pioneer air racer.

Aviation is proof, that given the will, we have the capacity to achieve the impossible.

—Captain Edward "Eddie" Rickenbacker.

Earthbound souls know only the underside of the atmosphere in which they live . . . but go higher—above the dust and water vapor—and the sky turns dark until one can see the stars at noon.

—Jacqueline Cochran, The Stars at Noon.

We thought humble and proud at the same time, all at once in love again with this painful bittersweet lovely thing called flight.

—Richard Bach, A Gift of Wings.

Flying prevails whenever a man and his airplane are put to a test of maximum performance.

—Richard Bach, A Gift of Wings.

Fighter pilot is an attitude. It is cockiness. It is aggressiveness. It is self-confidence. It is a streak of rebelliousness, and it is competitiveness. But there's something else—there's a spark. There's a desire to be good. To do well; in the eyes of your peers, and in your own mind.

I think it is love of that blue vault of sky that becomes your playground if, and only if, you are a fighter pilot. You don't understand it if you fly from A to B in straight and level, and merely climb and descend. You're moving through the basement of that bolt of blue.

A fighter pilot is a man in love with flying. A fighter pilot sees not a cloud but beauty. Not the ground but something remote from him, something that he doesn't belong to as long as he is airborne. He's a man who wants to be second-best to no one.

—Brigadier General Robin Olds, USAF, television interview.

Why fly? Simple. I'm not happy unless there's some room between me and the ground.

—Richard Bach, A Gift of Wings.

Flying is like sex—I've never had all I wanted but occasionally I've had all I could stand.

—Stephen Coonts, The Cannibal Queen.

But what I could never tell of was the beauty and exaltation of flying itself. Above the haze layer with the sun behind you or sinking ahead, alone in an open cockpit, there is nothing and everything to see. The upper surface of the haze stretches on like an endless desert, featureless and flat, and empty to the horizon. It seems your world alone. Threading one's way through the great piles of summer cumulus that hang over the plains, the patches of ground that show far far below are for earthbound folk, and the cloud shapes are sculptured just for you. The flash of rain, the shining rainbow riding completely around the plane, the lift over mountain ridges, the steady, pure air at dawn take-offs. . . . It was so alive and rich a life that any other conceivable choice seemed dull, prosaic, and humdrum.

—*Dean Smith,* By the Seat of My Pants.

Buddy of mine once told me that he'd rather fly a jet than kiss his girl. Said it gave him more of a kick.

—*Jerry Connell, in the movie* Air Cadet, *1951.*

Flight is romance—not in the sense of sexual attraction, but as an experience that enriches life.

—*Stephen Coonts,* The Cannibal Queen.

I have often said that the lure of flying is the lure of beauty. That the reasons flyers fly, whether they know it or not, is the aesthetic appeal of flying.

—*Amelia Earhart.*

But flies an eagle flight, bold and forth on, Leaving no tract behind.

—*William Shakespeare,* Timon of Athens, *Act I. Scene 1.*

Flyers have a sense of adventures yet to come, instead of dimly recalling adventures of long ago as the only moments in which they truly lived.

—*Richard Bach,* A Gift of Wings.

In our dreams we are able to fly . . . and that is a remembering of how we were meant to be.

—*Madeleine L'Engle,* Walking on Water.

In order to invent the airplane you must have at least a thousand years' experience dreaming of angels.

—*Arnold Rockman.*

Deftly they opened the brain of a child, and it was full of flying dreams.

—*Stanley Kunitz,* My Surgeons.

For a plane to fly well, it must be beautiful.

—*Marcel Dassault.*

Someday I would like to stand on the moon, look down through a quarter of a million miles of space and say, "There certainly is a beautiful earth out tonight."

—*Lieutenant Colonel William H. Rankin,* The Man Who Rode the Thunder.

I watched him strap on his harness and helmet, climb into the cockpit and, minutes later, a black dot falls off the wing two thousand feet above our field. At almost the same instant, a white streak behind him flowered out into the delicate wavering muslin of a parachute—a few gossamer yards grasping onto air and suspending below them, with invisible threads, a human life, and man who by stitches, cloth, and cord, had made himself a god of the sky for those immortal moments.

A day or two later, when I decided that I too must pass through the experience of a parachute jump, life rose to a higher level, to a sort of exhilarated calmness. The thought of crawling out onto the struts and wires hundreds of feet above the earth, and then giving up even that tenuous hold of safety and of substance, left me a feeling of anticipation mixed with dread, of confidence restrained by caution, of courage salted through with fear. How tightly should one hold onto life? How loosely give it rein? What gain was there for such a risk? I would have to pay in money for hurling my body into space. There would be no crowd to watch and applaud my landing. Nor was there any scientific objective to be gained. No, there was deeper reason for wanting to jump, a desire I could not explain.

It was that quality that led me into aviation in the first place—it was a love of the air and sky and flying, the lure of adventure, the appreciation of beauty. It lay beyond the descriptive words of man—where immortality is touched through danger, where life meets death on equal plane; where man is more than man, and existence both supreme and valueless at the same instant.

—*Charles A. Lindbergh.*

Fly, dotard, fly!
With thy wise dreams and fables of the sky.

—*Alexander Pope, The* Odyssey of Homer, *Book II.*

Flying is an act of conquest, of defeating the most basic and powerful forces of nature. It unites the violent rage and brute power of jet engines with the infinitesimal tolerances of the cockpit. Airlines take their measurements from the ton to the milligram, from the mile to the millimeter, endowing any careless move—an engine setting, a flap position, a training failure—with the power to wipe out hundreds of lives.

—*Thomas Petzinger, Jr., beginning of the Prologue to* Hard Landing.

The great bird will take its first flight . . . filling the world with amazement and all records with its fame, and it will bring eternal glory to the nest where it was born.

—*Leonardo da Vinci.*

A sky as pure as water bathed the stars and brought them out.

—*Antoine de Saint-Exupéry, first sentence of* Southern Mail, *1929.*

It is not enough to just ride this earth. You have to aim higher, try to take off, even fly. It is our duty.

—*José Yacopi, Argentine luthier.*

The higher we soar, the smaller we appear to those who cannot fly.

—Friedrich Wilhelm Nietzsche.

Flying might not be all plain sailing, but the fun of it is worth the price.

—Amelia Earhart.

Take possession of the air, submit the elements, penetrate the last redoubts of nature, make space retreat, make death retreat.

—Romain Rolland, 1912.

A man can criticize a pilot for flying into a mountainside in fog, but I would rather by far die on a mountainside than in bed. What sort of man would live where there is no daring? Is life itself so dear that we should blame one for dying in adventure? Is there a better way to die?

—Charles A. Lindbergh.

Flight is the only truly new sensation that men have achieved in modern history.

—James Dickey, The New York Times Book Review, *July 15, 1979.*

How many more years I shall be able to work on the problem I do not know; I hope, as long as I live. There can be no thought of finishing, for "aiming at the stars" both literally and figuratively, is a problem to occupy generations, so that no matter how much progress one makes, there is always the thrill of just beginning.

—Robert H. Goddard, in a 1932 letter to H. G. Wells.

Live thy life as it were spoil and pluck the joys that fly.

—Martial, Epigrams, *A.D. 86.*

Aviation will give new nourishment to the religious sprit of mankind. It will add airspace to those other great heighteners of the cosmic mood: the wood, the sea, the desert.

—Christian Morgenstern.

The helicopter has become the most universal vehicle ever created and used by man. It approaches closer than any other to fulfillment of mankind's ancient dreams of the flying horse and the magic carpet.

—Igor I. Sikorsky, comment on the 20th anniversary of the helicopter's first flight, September 13, 1959.

Many wonderful inventions have surprised us during the course of the last century and the beginning of this one. But most were completely unexpected and were not part of the old baggage of dreams that humanity carries with it. Who had ever dreamed of steamships, railroads, or electric light? We welcomed all these improvements with astonished pleasure; but they did not correspond to an expectation of our spirit or a hope as old as we are: to overcome gravity, to tear ourselves away from the earth, to become lighter, to fly away, to take possession of the immense aerial kingdom; to enter the universe of the Gods, to become Gods ourselves.

—Jerome Tharaud, Dans le ciel des dieux,
in Les Grandes Conferences de l'aviation: Recits et souvenirs, *1934.*

Predictions

There shall be wings! If the accomplishment be not for me, 'tis for some other. The spirit cannot die; and man, who shall know all and shall have wings . . .

—*Leonardo da Vinci.*

First, by the figurations of art there be made instruments of navigation without men to row them, as great ships to brooke the sea, only with one man to steer them, and they shall sail far more swiftly than if they were full of men; also chariots that shall move with unspeakable force without any living creature to stir them. Likewise an instrument may be made to fly withall if one sits in the midst of the instrument, and do turn an engine, by which the wings, being artificially composed, may beat the air after the manner of a flying bird.

—*Roger Bacon, thirteenth-century Franciscan friar.*

A bird is an instrument working according to mathematical law, which instrument is within the capacity of man to reproduce in all its movements.

—*Leonardo da Vinci, 1505.*

Flying would give such occasions for intrigues as people cannot meet with who have nothing but legs to carry them. You should have a couple of lovers make a midnight assignation upon the top of the monument, and see the cupola of St. Paul's covered with both sexes like the outside of a pigeon-house. Nothing would he more frequent than to see a beau flying in at a garret window, or a gallant giving chase to his mistress, like a hawk after a lark. The poor husband could not dream what was doing over his head. If he were jealous, indeed, he might clip his wife's wings, but what would this avail when there were flocks of whore-masters perpetually hovering over his house?

—*Joseph Addison*, The Guardian, *July 20, 1713.*

We have already begun to fly; several persons, here and there, have found the secret to fitting wings to themselves, of setting them in motion, so that they are held up in the air and are carried across streams . . . the art of flying is only just being born; it will be perfected, and some day we will go as far as the moon.

—*Bernard Le Bovier de Fontenelle (1657–1757).*

At sea let the British their neighbors defy—The French shall have frigates to traverse the sky.

—*Philip Freneau*, The Progress of Balloons, *1784.*

Soon shall thy arm, unconquer'd steam! afar
Drag the slow barge, or drive the rapid car;
Or on wide-waving wings expanded bear
The flying chariot through the field of air.

—*Erasmus Darwin, (1731–1802)*, The Botanic Garden, *Part I, Canto I.*

I confess that in 1901, I said to my brother Orville that man would not fly for fifty years . . . Ever since, I have distrusted myself and avoided all predictions.

—*Wilbur Wright, 1908.*

Engines of war have long since reached their limits, and I see no further hope of any improvement in the art.

—Julius Frontinus, A.D. 90.

But whether [the source of lift] be a rising current or something else, it is as well able to support a flying machine as a bird, if man once learns the art of utilizing it.

—Wilbur Wright, September 18, 1901.

He that can swim needs not despair to fly; to swim is to fly in a grosser fluid, and to fly is to swim in a subtler. We are only to proportion our power of resistance to the different density of matter through which we are to pass. You will be necessarily upborne by the air if you can renew any impulse upon it faster than the air can recede from the pressure . . . The labor of rising from the ground will be great, . . . but as we mount higher, the earth's attraction, and the body's gravity, will be gradually diminished till we arrive at a region where the man will float in the air without any tendency to fall.

—Dr. Samuel Johnson, The History of Rasselas, Prince of Abyssinia, *1759.*

Anyone who should see in the sky such a globe should be aware that, far from being an alarming phenomenon, it is only a machine made of taffetas or light canvas covered with paper, that cannot possibly cause any harm, and which will someday prove serviceable to the wants of society.

—French government proclamation issued to allay public alarm about balloon flights, 1784.

What is the use of a new-born infant?

—Benjamin Franklin, when asked what was the use of a balloon, while he was the American plenipotentiary to France in the early 1780s.

Bishop Wilkins prophesied that the time would come when gentlemen, when they were to go on a journey, would call for their wings as regularly as they call for their boots.

—Maria Edgeworth, Essay on Irish Bulls, *1802.*

I may be expediting the attainment of an object that will in time be found of great importance to mankind; so much so, that a new era in society will commence from the moment that aerial navigation is familiarly realised. . . . I feel perfectly confident, however, that this noble art will soon be brought home to man's convenience, and that we shall be able to transport ourselves and our families, and their goods and chattels, more securely by air than by water, and with a velocity of from 20 to 100 miles per hour.

—Sir George Cayley, 1809.

Darius was clearly of the opinon
That the air is also man's dominion
And that with paddle or fin or pinion,
We soon or late shall navigate
The azure as now we sail the sea.

—J. T. Trowbridge, Darius Green and His Flying Machine, *1869.*

Sir, Your letter of the 15th is received, but Age has long since obliged me to withhold my mind from Speculations of the difficulty of those of your letter, that there are means of artificial buoyancy by which man may be supported in the Air, the Balloon has proved, and that means of directing it may be discovered is against no law of Nature and is therefore possible as in the case of Birds, but to do this by mechanical means alone in a medium so rare and unassisting as air must have the aid of some principal not yet generally known. However, I can really give no opinion understandingly on the subject and with more good will than confidence wish to you success.

—*Thomas Jefferson, April 27, 1822.*

For I dipt into the future, far as human eye could see,
Saw the vision of the world, and all the wonder that would be;
Saw the heavens fill with commerce, Argosies of magic sails,
Pilots of the purple twilight, dropping down with costly bales;
Heard the heavens fill with shouting, and there rain'd a ghastly dew,
From the nations' airy navies grappling in the central blue.

—*Alfred, Lord Tennyson,* Locksley Hall, *1842.*

I suppose we shall soon travel by air-vessels; make air instead of sea voyages; and at length find our way to the moon, in spite of the want of atmosphere.

—*Lord Byron, 1882.*

And then, the Earth being small, mankind will migrate into space, and will cross the airless Saharas which separate planet from planet and sun from sun. The Earth will become a Holy Land which will be visited by pilgrims from all the quarters of the Universe. Finally, men will master the forces of Nature; they will become themselves architects of systems, manufacturers of worlds.

—*Winwood Reade,* The Martyrdom of Man, *1872.*

So, may it be; let us hope that the advent of a successful flying machine, now only dimly foreseen and nevertheless thought to be possible, will bring nothing but good into the world; that it shall abridge distance, make all parts of the globe accessible, bring men into closer relation with each other, advance civilisation, and hasten the promised era in which there shall be nothing but peace and good will among all men.

—*Octave Chanute, 1894.*

Heavier-than-air flying machines are impossible.

—*Lord Kelvin, President, Royal Society, 1895.*

Everything that can be invented has been invented.

—*Charles H. Duell, Commissioner, U.S. Office of Patents, 1899.*

Well, gentlemen, do you believe in the possibility of aerial locomotion by machines heavier than air? . . . You ask yourselves doubtless if this apparatus, so marvellously adapted for aerial locomotion, is susceptible of receiving greater speed. It is not worth while to conquer space if we cannot devour it. I wanted the air to be a solid support to me, and it is. I saw that to struggle against the wind I must be stronger than the wind, and I am. I had no need of sails to drive me, nor oars nor wheels to push me, nor rails to give me a faster road. Air is what I wanted, that was all. Air surrounds me as water surrounds the submarine boat, and in it my propellers act like the screws of a steamer. That is how I solved the problem of aviation. That is what a balloon will never do, nor will any machine that is lighter than air.

—*Jules Verne, 1886.*

For some years I have been afflicted with the belief that flight is possible to man. The disease has increased in severity and I feel that it will soon cost me an increased amount of money, if not my life.

—*Wilbur Wright, beginning of his first letter to Octave Chanute, May 13, 1900.*

I am intending to start out in a few days for a trip to the coast of North Carolina . . . for the purpose of making some experiments with a flying machine. It is my belief that flight is possible, and while I am taking up the investigation for pleasure rather than profit, I think there is a slight possibility of achieving fame and fortune from it.

—*Wilbur Wright, 1900.*

The demonstration that no possible combination of known substances, known forms of machinery and known forms of force, can be united in a practical machine by which man shall fly long distances through the air, seems to the writer as complete as it is possible for the demonstration of any physical fact to be.

—Simon Newcomb, 1901.

Few people who know of the work of Langley, Lilienthal, Pilcher, Maxim and Chanute but will be inclined to believe that long before the year 2000 A.D., and very probably before 1950, a successful aeroplane will have soared and come home safe and sound.

—H. G. Wells, 1901.

Flight by machines heavier than air is unpractical and insignificant, if not utterly impossible.

—Simon Newcomb, 1902.

The example of the bird does not prove that man can fly. Imagine the proud possessor of the aeroplane darting through the air at a speed of several hundred feet per second. It is the speed alone that sustains him. How is he ever going to stop?

—Simon Newcomb, in The Independent, *October 22, 1903.*

We are still far from the ultimate goal, and it would seem as if years of constant work and study by experts, together with the expenditure of thousands of dollars, would still be necessary before we can hope to produce an apparatus of practical utility on these lines.

—The U.S. War Department, in its final report on the Langley Project, 1903.

I believe the new machine of the Wrights to be the most promising attempt at flight that has yet been made.

—Octave Chanute, November 23, 1903.

We hope that Professor Langley will not put his substantial greatness as a scientist in further peril by continuing to waste his time and the money involved in further airship experiments. Life is short, and he is capable of services to humanity incomparably greater than can be expected to result from trying to fly. . . . For students and investigators of the Langley type, there are more useful employments.

—The New York Times, *editorial page of December 10, 1903, 1 week before the flight at Kitty Hawk.*

I remember how my comrades used to tease me at our game of "Pigeon flies!" All the children gather round a table and the leader calls out "Pigeon flies! Hen flies! Crow flies! Bee flies!"and so on; and at each call we were supposed to raise our fingers. Sometimes, however, he would call out: "Dog flies! Fox flies!"or some other like impossibility, to catch us. If any one raised a finger, he was made to pay a forfeit. Now my playmates never failed to wink and smile mockingly at me when one of them called "Man flies!" for at the word I would always lift my finger very high, as a sign of absolute conviction; and I refused with energy to pay the forfeit. The more they laughed at me, the happier I was, hoping that some day the laugh would be on my side.

—Alberto Santos-Dumont, My Air-Ships, *New York, The Century Company, 1904.*

It seems to me that the conquest of the air is the only major task for our generation.

—T. E. Lawrence.

This fellow Charles Lindbergh will never make it. He's doomed.

—Harry Guggenheim, millionaire aviation enthusiast.

The aeroplane will never fly.

—Lord Haldane, The British Minister of War, 1907 (yes, 1907).

It is a bare possibility that a one-man machine without a float and favored by a wind say of 15 miles an hour might succeed in getting across the Atlantic. But such an attempt would be the height of folly. When one comes to increase the size of the craft, the possibility rapidly fades away. This is because of the difficulties of carrying sufficient fuel. It will readily be seen, therefore, why the Atlantic flight is out of the question.

—Orville Wright, circa 1908.

No flying machine will ever fly from New York to Paris . . . [because] no known motor can run at the requisite speed for four days without stopping.

—*Orville Wright, circa 1908.*

The popular mind often pictures gigantic flying machines speeding across the Atlantic, carrying innumerable passengers. It seems safe to say that such ideas must be wholly visionary. Even if such a machine could get across with one or two passengers, it would be prohibitive to any but the capitalist who could own his own yacht.

—*William Pickering, Harvard astronomer, circa 1913.*

Oh well, I suppose lots of people will do it now.

—*Arthur Whitten Brown to Captain John Alcock after they crash-landed in a bog at Cliften, Ireland, after completing the first transatlantic flight, 1919.*

Scientific investigation into the possibilities [of jet propulsion] has given no indication that this method can be a serious competitor to the airscrew-engine combination.

—*The British Undersecretary of State for Air, 1934.*

In its present state, and even considering the improvements possible when adopting the higher temperatures proposed for the immediate future, the gas turbine engine could hardly be considered a feasible application to airplanes mainly because of the difficulty in complying with the stringent weight requirements imposed by aeronautics.

The present internal combustion engine equipment used in airplanes weighs about 1.1 pounds per horsepower, and to approach such a figure with a gas turbine seems beyond the realm of possibility with existing materials.

—The Committee on Gas Turbines appointed by The National Academy of Sciences, June 10, 1940.

Reflecting on this, Frank Whittle remarked, "Good thing I was too stupid to know this."

We do not consider that aeroplanes will be of any possible use for war purposes.

—The British Secretary of State for War, 1910.

Airplanes are interesting toys but of no military value.

—Marshal Ferdinand Foch, professor of strategy, École Supériure de Guerre, 1911.

Thank God men cannot as yet fly and lay waste the sky as well as the earth!

—Henry David Thoreau (1817–1862).

The air around London and other large cities will be darkened by the flight of aeroplanes. . . . They are not mere dreamers who hold that the time is at hand when air power will be an even more important thing than sea power.

—The Daily Mail *newspaper, 1906.*

No place is safe—no place is at peace. There is no place where a woman and her daughter can hide and be at peace. The war comes through the air, bombs drop in the night. Quiet people go out in the morning, and see airfleets passing overhead—dripping death—dripping death.

—H. G. Wells, The War in the Air, 1908.

The way to fly is to go straight up . . . Such a machine [the helicopter] will never compete with the aeroplane, though it will have specialized uses, and in these it will surpass the aeroplane. The fact that you can land at your front door is the reason you can't carry heavy loads efficiently.

—Emile Berliner, 1948.

Someone asked the master about the principles [tao] of mounting to dangerous heights and traveling into the vast inane. The Master said, "Some have made flying cars [fei chhe] with wood from the inner part of the jujube tree, using ox-leather [straps] fastened to returning blades so as to set the machine in motion [huan chien i yin chhi chi]."

—Pao Phu Tau, fourth century A.D., earliest description of a helicopter?

I have discovered that a screw-shaped device such as this, if it is well made from starched linen, will rise in the air if turned quickly.

—Leonardo da Vinci, Codice Atlantico, describing his helical air screw, 1480.

We soon saw that the helicopter had no future, and dropped it. The helicopter does with great labor only what the balloon does without labor, and is no more fitted than the balloon for rapid horizontal flight. If its engine stops, it must fall with deathly violence, for it can neither glide like the aeroplane or float like the balloon. The helicopter is much easier to design than the aeroplane, but is worthless when done.

—Wilbur Wright, 1909.

'Tis likely enough that there may be means invented of journeying to the moon; and how happy they shall be that are first successful in this attempt.

—Dr John Wilkins, A Discourse Concerning a New World and Another Planet, *1640.*

Men might as well project a voyage to the Moon as attempt to employ steam navigation against the stormy North Atlantic Ocean.

—Dr. Dionysus Lardner, professor of natural philosophy and astronomy, University College, London, 1838.

This foolish idea of shooting at the moon is an example of the absurd length to which vicious specialisation will carry scientists working in thought-tight compartments.

—Professor Bickerton, speech delivered to the British Association for the Advancement of Science, 1929.

Our descendants will certainly attempt journeys to other members of the solar system. . . . By 2030 the first preparations for the first attempt to reach Mars may perhaps be under consideration. The hardy individuals who form the personnel of the expedition will be sent forth in a machine propelled like a rocket.

—Lord Birkenhead, 1930.

Earth is the cradle of mankind, but man cannot live in the cradle forever.

—Konstantin E. Tsiolkovsky, early Russian rocketry theorist.

Mankind will not remain on Earth forever, but in its quest for light and space will at first timidly penetrate beyond the confines of the atmosphere, and later will conquer for itself all the space near the Sun.

—Konstantin E. Tsiolkovsky.

Man will never reach the moon regardless of all future scientific advances.

—Dr. Lee De Forest.

[Airmail was] an impractical sort of fad, and had no place in the serious job of postal transportation.

—Colonel Paul Henderson, U.S. second assistant postmaster general, 1919.

[Before man reaches the moon] your mail will be delivered within hours from New York to California, to England, to India or to Australia by guided missiles. . . . We stand on the threshold of rocket mail.

—Arthur E. Summerfield, Associated Press wire report, January 23, 1959.

The concept is interesting and well-formed, but in order to earn better than a "C," the idea must be feasible.

—A Yale University management professor in response to Fred Smith's paper proposing reliable overnight delivery service. Fred Smith later started FedEx.

You fucking academic eggheads! You don't know shit. You can't deregulate this industry. You're going to wreck it. You don't know a goddamn thing!

—Robert L. Crandall, addressing a Senate lawyer prior to airline deregulation, 1977.

Deregulation will be the greatest thing to happen to the airlines since the jet engine.

—Richard J. Ferris, CEO of United Airlines, 1976.

United has little to fear from numerous small competitors. We should be able to compete effectively by advertising our size, dependability, and experience, and by matching or beating their promotional tactics. . . . In a free environment, we would be able to flex our marketing muscles a bit and should not fear the threat of being nibbled to death by little operators.

—Richard J. Ferris, CEO of United Airlines, 1976.

Total deregulation would allow anybody to fly any route, a situation that is unlikely ever to occur.

—The New York Times Magazine, *May 9, 1976.*

No one expects Braniff to go broke. No major U.S. carrier ever has.

—The Wall Street Journal, *July 30, 1980.*

But the airplane's potential would not—in fact, could not—be realized by a community of businessmen acting alone. The Federal Government would stand at their side, becoming, in effect, civil aviation's indispensable partner. The partnership flourishes to this day.

—*Nick A. Komons, FAA historian,* Bonfires to Beacons, *1977.*

The creative conquest of space will serve as a wonderful substitute for war.

—*James S. McDonnell,* Time, *March 31, 1967.*

I think that it is much more likely that the reports of flying saucers are the results of the known irrational characteristics of terrestrial intelligence than of the unknown rational efforts of extra-terrestrial intelligence.

—*Dr. Richard Feynman.*

Our journeys to the stars will be made on spaceships created by determined, hardworking scientists and engineers applying the principles of science, not aboard flying saucers piloted by little gray aliens from some other dimension.

—*Robert A. Baker,* The Aliens Among Us: Hypnotic Regression Revisited.

The phenomenon of UFOs does exist, and it must be treated seriously.

—*Mikhail Gorbachev,* Soviet Youth, *May 4, 1990.*

As a method of sending a missile to the higher, and even to the highest parts of the earth's atmospheric envelope, Professor Goddard's rocket is a practicable and therefore promising device. It is when one considers the multiple-charge rocket as a traveler to the moon that one begins to doubt . . . for after the rocket quits our air and really starts on its journey, its flight would be neither accelerated nor maintained by the explosion of the charges it then might have left. Professor Goddard, with his "chair" in Clark College and countenancing of the Smithsonian Institution, does not know the relation of action to re-action, and of the need to have something better than a vacuum against which to react . . . Of course he only seems to lack the knowledge ladled out daily in high schools.

—The New York Times, *editorial, 1920.*

It is difficult to say what is impossible, for the dream of yesterday is the hope of today and reality of tomorrow.

—*Robert H. Goddard.*

The greatest advance in aviation since the Wright Brothers.

—The New York Times, *1961. This a much overused phrase, that here describes the start of the Eastern air shuttle between New York and Washington.*

Centuries hence, when current social and political problems may seem as remote as the problems of the Thirty Years' War are to us, our age may be remembered chiefly for one fact: It was the time when the inhabitants of the earth first made contact with the vast cosmos in which their small planet is embedded.

—*Carl Sagan.*

Flight out of the atmosphere is a simple thing to do and should have been available to the public twenty years ago. Ten years from now, we will have space tourism where you will be able to see the black sky and the curvature of the earth. It will be the most exciting roller coaster ride you can buy.

—*Burt Rutan, in an interview with* Design News, *1996.*

Market studies suggest space tourism—a rubbernecker's trip to earth orbit—is likely to draw 50,000 passengers a year if the ticket can be pushed below $25,000. That's what tens of thousands of people spend each year on competing trips, such as round-the-world cruises on luxury liners and adventure tours to Antarctica or Mount Everest.

—*G. Harry Stine,* Barron's, *editorial commentary, October 21, 1996.*

One day the stars will be as familiar to each man as the landmarks, the curves, and the hills on the road that leads to his door, and one day that will be an airborne life.

—Beryl Markham, West With the Night.

There are no practical alternatives to air transportation.

—Daniel S. Goldin, NASA administrator, March 20, 1997.

The past is but the beginning of a beginning, and all that is and has been is but the twilight of the dawn.

—H. G. Wells, The Discovery of the Future, *1901.*

First Flights

With a short dash down the runway, the machine lifted into the air and was flying. It was only a flight of twelve seconds, and it was an uncertain, wavy, creeping sort of flight at best; but it was a real flight at last and not a glide.

—Orville Wright, first flight of a heavier-than-air aircraft.

The course of the flight up and down was exceedingly erratic, partly due to the irregularity of the air, and partly to lack of experience in handling this machine.

—Orville Wright.

Those who understand the real significance of the conditions under which we worked will be surprised rather at the length than the shortness of the flights made with an unfamiliar machine after less than one minute's practice. The machine possesses greater capacity of being controlled than any of our former machines.

—Wilbur Wright.

Success four flights Thursday morning all against twenty one mile wind started from Level with engine power alone average speed through air thirty one miles longest 57 seconds inform Press home Christmas.

—Orevelle Wright.

This first telegram home had two transcription errors.
It should have read "59 seconds" and Orville was misspelled.

I was surprised at the silence and the absence of movement which our departure caused among the spectators, and believed them to be astonished and perhaps awed at the strange spectacle; they might well have reassured themselves. I was still gazing when M. Rozier cried to me—"You are doing nothing, and the balloon is scarcely rising a fathom."

"Pardon me," I answered, as I placed a bundle of straw upon the fire and slightly stirred it. Then I turned quickly but already we had passed out of sight of La Muette. Astonished I cast a glance towards the river. I perceived the confluence of the Oise. And naming the principal bends of the river by the places nearest them, I cried, "Passy, St. Germain, St. Denis, Sevres!"

"If you look at the river in that fashion you will be likely to bathe in it soon," cried Rozier. "Some fire, my dear friend, some fire!"

—*Marquis D'Arlandes, first flight of a hot air balloon, November 21, 1783.*

Nothing will ever equal that moment of exhilaration which filled my whole being when I felt myself flying away from the earth. It was not mere pleasure; it was perfect bliss . . .

—*Professor Jacques Alexandre Cesare Charles,*
first free flight in a manned hydrogen balloon, December 1, 1783. Note: The exact
adjective used by Professor Charles to describe his emotions in French is not "exhilaration"
but "hilarite," which can be translated as ecstasy, exhilaration, joy, and/or excitement.

When it first turned that circle, and came near the starting point, I was right in front of it; and said then, and I believe still, it was . . . the grandest sight of my life. Imagine a locomotive that has left its track, and is climbing up in the air right toward you—a locomotive without any wheels . . . but with white wings instead . . . a locomotive made of aluminum. Well, now, imagine this white locomotive, with wings that spread 20 feet each way, coming right toward you with a tremendous flap of its propellers, and you will have something like what I saw. The younger brother bade me move to one side for fear it might come down suddenly; but I tell you, friends, the sensation that one feels in such a crisis is something hard to describe.

—Amos Ives Root, witnessing the first time a heavier-than-air flying machine flew a complete circle, in the journal Gleanings in Bee Culture, *1904.*

I headed for this white mountain, but was caught in the wind and the mist . . . I followed the cliff from north to south, but the wind, against which I was fighting, got even stronger. A break in the coast appeared to my right, just before Dover Castle. I was madly happy. I headed for it. I rushed for it. I was above ground!

—Louis Blériot, first flight across the English Channel.

The hardships and perils of the past month were forgotten in the excitement of the present. We shook hands with one another, our hearts swelling with those emotions invoked by achievement and the glamour of the moment. It was, and will be, perhaps the supreme hour of our lives.

—Sir Ross Smith, K.B.E., first flight from London to Australia.

Demonstrated publicly at the Cuatro Vientos airport in Spain, the craft amazed and fascinated the whole aeronautical world. It was safe. Once . . . it climbed too steeply and lost all its forward motion, which, for the conventional aeroplane, would have meant plummeting to earth. This did not occur.

—Colonel H. F. Gregory, USAAF, witnessing the first autogiro.

These phantoms speak with human voices . . . able to vanish or appear at will, to pass in and out through the walls of the fuselage as though no walls were there . . . familiar voices, conversing and advising on my flight, discussing problems of my navigation, reassuring me, giving me messages of importance unattainable in ordinary life.

—Charles A. Lindbergh, first solo flight across the Atlantic.

Where am I?

—Charles A. Lindbergh, upon arrival in Paris.

I was a passenger on the journey . . . just a passenger. Everything that was done to bring us across was done by Wilmer Stultz and Slim Gordon. Any praise I can give them they ought to have . . . I do not believe that women lack the stamina to do a solo trip across the Atlantic, but it would be a matter of learning the arts of flying by instruments only, an art which few men pilots know perfectly now . . .

—Amelia Earhart, first flight of a woman across the Atlantic.

I'm Douglas Corrigan. Just got in from New York, where am I? . . . I intended to fly to California.

—Douglas "Wrong Way" Corrigan, upon arrival in Ireland after his unapproved solo transatlantic flight. He maintained he had "compass troubles."

Where am I?

In Gallegher's pasture . . . have you come far?

From America.

—Amelia Earhart, first solo flight by a woman across the Atlantic, upon arrival in an open field near Londonderry, Northern Ireland.

Somewhat to my dismay Everest bore that immense snow plume which means a mighty wind tearing across the summit, lifting clouds of powered snow and driving it with blizzard force eastward. Up went the machine into a sky of indescribable blue till we came on a level with the great peak itself. This astonishing picture of Everest, its plume now gradually lessening, its tremendous southern cliffs flanked by Makalu, was a sight which must remain in the mind all the years of one's life.

—Lieutenant Colonel L. V. Stewart Blacker, first flight over Everest.

I happened to be the man on the spot, but any of the rest of the fellows would have done what I did.

—Jack Knight, first night mail flight, which was part of a record-setting transcontinental airmail relay.

This machine was a failure to the extent that it could not fly. In other respects it was a very important and necessary stepping stone.

—*Igor I. Sikorsky, about the first helicopter, built 1909.*

Apart from a few tricky minutes in low cloud near the North Downs the journey over Folkestone and Boulogne down to Beauvais was uneventful but wet and hardly ever over 200 feet above ground . . . we eventually landed at Le Bourget at 10:15 A.M. In those days the airfield consisted of several canvas hangars, some wooden sheds and a lot of mud.

—*Jerry Shaw, first flight of a paying passenger from England to France.*

For the first time I was flying by jet propulsion. No engine vibrations. No torque and no lashing sound of the propeller. Accompanied by a whistling sound, my jet shot through the air. Later when asked what it felt like, I said, "It felt as though angels were pushing."

—*Lieutenant Général Adolf Galland, his first jet aircraft flight.*

Leveling off at 42,000 feet, I had thirty percent of my fuel, so I turned on rocket chamber three and immediately reached .96 Mach. I noticed that the faster I got, the smoother the ride. Suddenly the Mach needle began to fluctuate. It went up to .965 Mach—then tipped right off the scale . . . We were flying supersonic. And it was a smooth as a baby's bottom; Grandma could be sitting up there sipping lemonade.

—*General Charles "Chuck" Yeager, first supersonic flight.*

. . . Your seat pushed you firmly in the back. Even then there is none of the shuddering brazen bellow of the high-powered piston engine . . . Combined with a seemingly uncanny lack of vibration, this gives the impression almost of sailing through space, the engines with their glinting propeller discs utterly remote from the quiet security of this cabin.

—Derek Harvey, first turboprop airliner flight, the Vickers-Armstrong Viscount.

Millions wonder what it is like to travel in the Comet at 500 miles an hour eight miles above the earth. Paradoxically there is a sensation of being poised motionless in space. Because of the great height the scene below scarcely appears to move; because of the stability of the atmosphere the aircraft remains rock-steady . . . One arrives over distant landmarks in an incredibly short time but without the sense of having traveled. Speed does not enter into the picture. One doubts one's wristwatch.

—C. Martin Sharp, first jet airliner flight, the de Havilland Comet 1.

I tell you, we're going to be busy for a minute.

—Neil Armstrong, one of the first transmissions from Tranquillity Base.

Our A320 behaved even better than expected—it is both delightfully responsive and reassuringly stable to fly, qualities which fly-by-wire brings together for the first time in an airliner. Never before have we enjoyed a first flight so much, and we are confident that airline pilots will feel the same way.

—Pierre Baud, first flight of a fully fly-by-wire airliner.

Air Power

God would surely never allow such a machine to be successful, since it would cause much disturbance among the civil and political governments of mankind . . . no city would be proof against surprise . . . or ships that sail the sea. . . . Houses, fortresses, and cities could thus be destroyed, with the certainty that the airship would come to no harm, as the missiles could be thrown from a great height.

—Francesco de Lana de Terzide, Francesco of Brescia,
Italian Jesuit who was the first Westerner to write on the military uses of aerial attack, 1670.

When my brother and I built the first man-carrying flying machine we thought that we were introducing into the world an invention which would make further wars practically impossible.

—Orville Wright, 1917.

And where is the Prince who can afford to so cover his country with troops for its defense, as that ten thousand men descending from the clouds, might not in many places do an infinite deal of mischief, before a force could be brought to-gether to repel them?

—Benjamin Franklin.

Once the command of the air is obtained by one of the contended armies, the war must become a conflict between a seeing host and one that is blind.

—H. G. Wells.

We were once told that the aeroplane had "abolished frontiers." Actually it is only since the aeroplane became a serious weapon that frontiers have become definitely impassable.

—George Orwell.

The cavalry, in particular, were not friendly to the aeroplane, which it was believed, would frighten the horses.

—Sir Walter Raleigh.

Bombardment from the air is legitimate only when directed at a military objective, the destruction or injury of which would constitute a distinct military disadvantage to the belligerent.

—The Hague Convention of Jurists, 1923.

Above all, I shall see to it that the enemy will not be able to drop any bombs.

—Reichsmarschall Hermann Wilhelm Goering. German original:
"Vor allem werde ich dafür sorgen, dass der Feind keint Bomben werfen Kann."

The bomber will always get through.

—British Prime Minister Stanley Baldwin, 1932.

Believe me, Germany is unable to wage war.

—Former British Prime Minister David Lloyd George, 1934.

I wish for many reasons flying had never been invented.

—British Prime Minister Stanley Baldwin, on learning that Germany
had secretly built an air force in defiance of the Treaty of Versailles, 1935.

The military mind always imagines that the next war will be on the same lines as the last. That has never been the case and never will be. One of the great factors of the next war will be aircraft obviously. The potentialities of aircraft attack on a large scale are almost incalculable.

—Marshal of France Ferdinand Foch.

What General Weygand called the Battle of France is over. I expect the Battle of Britain is about to begin . . . Let us therefore brace ourselves to our duties, and so bear ourselves that, if the British Empire and its Commonwealth last for a thousand years, men will still say, "This was their finest hour."

—Sir Winston Churchill, June 18, 1940.

Never in the field of human conflict was so much owed by so many to so few.

—Sir Winston Churchill, House of Commons, August 20, 1940.
The Royal Air Force has been known as "the few" ever since.
This quote is often changed by writers and speakers, giving us material such as,
"Never . . . was so much owed by so few to so many," heard after the Falklands War.

Space in which to maneuver in the air, unlike fighting on land or sea, is practically unlimited, and . . . any number of airplanes operating defensively would seldom stop a determined enemy from getting through. Therefore the airplane was, and is, essentially an instrument of attack, not defense.

—Air Vice-Marshal James Edgar "Johnnie" Johnson," RAF.

Only air power can defeat air power. The actual elimination or even stalemating of an attacking air force can be achieved only by a superior air force.

—Major Alexander P. de Severskyd., USAAF.

Offense is the essence of air power.

—General Henry Harley "Hap" Arnold, USAAF.

How could they possibly be Japanese planes?

—Admiral Husband E. Kimmel.

No aircraft ever took and held ground.

—U.S. Marine Corps Manual.

Air power is indivisible. If you split it up into compartments, you merely pull it to pieces and destroy its greatest asset—its flexibility.

—Field Marshal Bernard Montgomery.

Because of its independence of surface limitations and its superior speed the airplane is the offensive weapon par excellence.

—General Giulio Douhet.

I have a mathematical certainty that the future will confirm my assertion that aerial warfare will be the most important element in future wars, and that in consequence not only will the importance of the Independent Air Force rapidly increase, but the importance of the army and navy will decrease in proportion.

—General Giulio Douhet, Command of the Air, *1921.*

The most important branch of aviation is pursuit, which fights for and gains control of the air.

—Brigadier General William "Billy" Mitchell, USAS.

In the development of air power, one has to look ahead and not backward and figure out what is going to happen, not too much what has happened.

—Brigadier General William "Billy" Mitchell, USAS.

[In World War I] air raids on both sides caused interruptions to production and transportation out of all proportion to the weight of bombs dropped.

—Edward Meade Earle.

The advent of air power, which can go straight to the vital centers and either neutralize or destroy them, has put a completely new complexion on the old system of making war. It is now realized that the hostile main army in the field is a false objective, and the real objectives are the vital centers.

—*Brigadier General William "Billy" Mitchell,* Skyways: A Book on Modern Aeronautics, *1930.*

I believe this plan [raiding RAF airfields] would have been very successful, but as a result of the Führer's speech about retribution, in which he asked that London be attacked immediately, I had to follow the other course. I wanted to attack the airfields first, thus creating a prerequisite for attacking London . . . I spoke with the Führer about my plans in order to try to have him agree I should attack the first ring of RAF airfields around London, but he insisted he wanted to have London itself attacked for political reasons, and also for retribution.

I considered the attacks on London useless, and I told the Führer again and again that inasmuch as I knew the English people as well as I did my own people, I could never force them to their knees by attacking London. We might be able to subdue the Dutch people by such measures but not the British.

—*Reichmarschall Hermann Wilhelm Goering, International Military Tribunal, Nuremberg, 1946.*

Strategic air assault is wasted if it is dissipated piecemeal in sporadic attacks between which the enemy has an opportunity to readjust defenses or recuperate.

—*General Henry Harley "Hap" Arnold, USAAF.*

The sky over London was glorious, ochre and madder, as though a dozen tropic suns were simultaneously setting round the horizon . . . Everywhere the shells sparkled like Christmas baubles.

—Evelyn Waugh.

From now on we shall bomb Germany on an ever-increasing scale, month by month, year by year, until the Nazi regime has either been exterminated by us or—better still—torn to pieces by the German people themselves.

—Sir Winston Churchill, July 14, 1941.

Against this pale, duck-egg blue and greyish mauve were silhouetted a number of small black shapes: all of them bombers, and all of them moving the same way. One hundred and thirty-four miles ahead, and directly in their path, stretched a crimson-red glow; Cologne was on fire. Already, only twenty-three minutes after the attack had started, Cologne was ablaze from end to end, and the main force of the attack was still to come.

—Group Captain Leonard Cheshire, VC, DSO.

If we had these rockets in 1939, we should never have had this war.

—Adolf Hitler, with regard to the V-1 rocket.

If we lose the war in the air we lose the war and lose it quickly.

—Field Marshal Bernard Montgomery.

Air power is like poker. A second-best hand is like none at all—it will cost you dough and win you nothing.

—George Kenney.

Air power alone does not guarantee America's security, but I believe it best exploits the nation's greatest asset—our technical skill.

—General Hoyt S. Vandenberg, second U. S. Air Force Chief of Staff.

Air battle is not decided in a few great clashes but over a long period of time when attrition and discouragement eventually cause one side to avoid the invading air force.

—Dale O. Smith.

And even if a semblance of order could be maintained and some work done, would not the sight of a single enemy airplane be enough to induce a formidable panic? Normal life would be unable to continue under the constant threat of death and imminent destruction.

—General Giulio Douhet, "Il dominio dell'aria," 1921.

The conviction of the justification of using even the most brutal weapons is always dependent on the presence of a fanatical belief in the necessity of the victory of a revolutionary new order on this globe.

—Adolf Hitler, Mein Kampf.

Bombing is often called "strategic"' when we hit the enemy, and "tactical" when he hits us, and is often difficult to know where one finishes and the other begins.

—*Air Vice-Marshal James Edgar "Johnnie" Johnson, RAF.*

Airpower has become predominant, both as a deterrent to war, and—in the eventuality of war—as the devastating force to destroy an enemy's potential and fatally undermine his will to wage war.

—*General Omar Bradley.*

Air Power is, above all, a psychological weapon—and only short-sighted soldiers, too battle-minded, underrate the importance of psychological factors in war.

—*B. H. Liddell-Hart.*

The future battle on the ground will be preceded by battle in the air. This will determine which of the contestants has to suffer operational and tactical disadvantages and be forced throughout the battle into adopting compromise solutions.

—*Field Marshal Erwin Rommel.*

Anyone who has to fight, even with the most modern weapons, against an enemy in complete command of the air, fights like a savage against modern European troops, under the same handicaps and with the same chances of success.

—*Field Marshal Erwin Rommel,* Rommel Papers, *1953.*

. . . then there was war in heaven. But it was not angels. It was that small golden zeppelin, like a long oval world, high up. It seemed as if the cosmic order were gone, as if there had come a new order, a new heavens above us: and as if the world in anger were trying to revoke it . . . So it seems ours cosmos is burst, burst at last, the stars and moon blown away, the envelope of the sky burst out, and a new cosmos appeared, with a long-ovate, gleaming central luminary, calm and drifting in a glow of light, like a new moon, with its light bursting in flashes on the earth, to burst away the earth also. So it is the end—our world is gone, and we are like dust in the air.

—*John Milton,* Paradise Lost.

Anyone who has to fight, even with the most modern weapons, against an enemy in complete command of the air, fights like a savage against modern European troops, under the same handicaps and with the same chances of success.

—*Field Marshal Erwin Rommel,* Rommel Papers, *1953.*

The first and absolute requirement of strategic air power in this war was control of the air in order to carry out sustained operations without prohibitive losses.

—*General Carl A. "Tooey" Spaatz.*

In order to assure an adequate national defense, it is necessary—and sufficient— to be in a position in case of war to conquer the command of the air.

—*General Giulio Douhet.*

A modern, autonomous, and thoroughly trained Air Force in being at all times will not alone be sufficient, but without it there can be no national security.

—General Henry Harley "Hap" Arnold, USAAF.

Air control can be established by superiority in numbers, by better employment, by better equipment, or by a combination of these factors.

—General Carl A. "Tooey" Spaatz.

A modern state is such a complex and interdependent fabric that it offers a target highly sensitive to a sudden and overwhelming blow from the air.

—B. H. Liddell-Hart.

We carried out many trials to try to find the answer to the fast, low-level intruder, but there is no adequate defense.

—Air Vice-Marshal James Edgar "Johnnie" Johnson, RAF.

The best way to defend the bombers is to catch the enemy before it is in position to attack. Catch them when they are taking off, or when they are climbing, or when they are forming up. Don't think you can defend the bomber by circling around him. It's good for the bombers morale, and bad for tactics.

—Brigadier General Robin Olds, USAF.

To have command of the air means to be able to cut an enemy's army and navy off from their bases of operation and nullify their chances of winning the war.

—General Giulio Douhet.

In the early stages of the fight Mr. Winston Churchill spoke with affectionate raillery of me and my "Chicks." He could have said nothing to make me more proud; every Chick was needed before the end.

—ACM Sir Hugh C. T. Dowding, dispatch to the Secretary of State for Air, August 20, 1941.

The most important thing is to have a flexible approach. . . . The truth is no one knows exactly what air fighting will be like in the future. We can't say anything will stay as it is, but we also can't be certain the future will conform to particular theories, which so often, between the wars, have proved wrong.

—Brigadier General Robin Olds, USAF.

The weapon where the man is sitting in is always superior against the other.

—Colonel Erich "Bubi" Hartmann, GAF.

"He who wants to protect everything, protects nothing" is one of the fundamental rules of defense.

—Lieutenant General Adolf Galland, Luftwaffe.

If we should have to fight, we should be prepared to do so from the neck up instead of from the neck down.

—General James "Jimmy" Doolittle.

To use a fighter as a fighter-bomber when the strength of the fighter arm is inadequate to achieve air superiority is putting the cart before the horse.

—Lieutenant General Adolf Galland, Luftwaffe.

The only proper defense is offense.

—Air Vice-Marshal James Edgar "Johnnie" Johnson, RAF.

It is not possible to seal an air space hermetically by defensive tactics.

—Air Vice-Marshal James Edgar "Johnnie" Johnson, RAF.

Superior technical achievements—used correctly both strategically and tactically —can beat any quantity numerically many times stronger yet technically inferior.

—Lieutenant General Adolf Galland, Luftwaffe.

Good airplanes are more important than superiority in numbers.

—Air Vice-Marshal James Edgar "Johnnie" Johnson, RAF.

Why don't we just buy one airplane and let the pilots take turns flying it?

—Calvin Coolidge, complaining about a War Department request to buy more aircraft.

We have the enemy surrounded. We are dug in and have overwhelming numbers. But enemy airpower is mauling us badly. We will have to withdraw.

—Japanese infantry commander, situation report to headquarters, Burma, World War II.

In our victory over Japan, airpower was unquestionably decisive. That the planned invasion of the Japanese Home islands was unnecessary is clear evidence that airpower has evolved into a force in war co-equal with land and sea power, decisive in its own right and worthy of the faith of its prophets.

—General Carl A. "Tooey" Spaatz., "Evolution of Air Power," Military Review, *1947.*

The function of the Army and Navy in any future war will be to support the dominant air arm.

—General James "Jimmy" Doolittle, in a speech at Georgetown University, 1949.

If we maintain our faith in God, love of freedom, and superior global air power, the future [of the United States] looks good.

—General Curtis Emerson LeMay, USAF.

But I have seen the science I worshiped, and the airplane I loved, destroying the civilization I expected them to serve.

—Charles A. Lindbergh, Time, *May 26, 1967.*

Combat

So it was that the war in the air began. Men rode upon the whirlwind that night and slew and fell like archangels. The sky rained heroes upon the astonished earth. Surely the last fights of mankind were the best. What was the heavy pounding of your Homeric swordsmen, what was the creaking charge of chariots, besides this swift rush, this crash, this giddy triumph, this headlong sweep to death?

—*H. G. Wells,* The World Set Free, *1914.*

Good flying never killed [an enemy] yet.

—*Major Edward "Mick" Mannock, RAF.*

There are only two types of aircraft—fighters and targets.

—*Doyle "Wahoo" Nicholson, USMC.*

Up there the world is divided into bastards and suckers. Make your choice.

—*Derek Robinson,* Piece of Cake.

I hate to shoot a Hun down without him seeing me, for although this method is in accordance with my doctrine, it is against what little sporting instincts I have left.

—*Major James Thomas Byford McCudden, VC, RFC, 1917.*

The first time I ever saw a jet, I shot it down.

—*General Charles "Chuck" Yeager, USAF, describing his first confrontation with an Me-262.*

Of all my accomplishments I may have achieved during the war, I am proudest of the fact that I never lost a wingman.

—Colonel Erich "Bubi" Hartmann, GAF, aka Karaya One,
world's leading ace, 352 victories in World War II.

It was my view that no kill was worth the life of a wingman. . . . Pilots in my unit who lost wingmen on this basis were prohibited from leading a [section]. They were made to fly as wingman, instead.

—Colonel Erich "Bubi" Hartmann, GAF.

There is a peculiar gratification on receiving congratulations from one's squadron for a victory in the air. It is worth more to a pilot than the applause of the whole outside world. It means that one has won the confidence of men who share the misgivings, the aspirations, the trials and the dangers of aeroplane fighting.

—Captain Edward V. "Eddie" Rickenbacker, USAS.

Nothing makes a man more aware of his capabilities and of his limitations than those moments when he must push aside all the familiar defenses of ego and vanity, and accept reality by staring, with the fear that is normal to a man in combat, into the face of Death.

—Major Robert S. Johnson, USAAF.

It is probable that future war will be conducted by a special class, the air force, as it was by the armored Knights of the Middle Ages.

—*Brigadier General William "Billy" Mitchell,* Winged Defense, *1924.*

The duty of the fighter pilot is to patrol his area of the sky, and shoot down any enemy fighters in that area. Anything else is rubbish.

—*Baron Manfred Rittmeister von Richthofen, "The Red Baron," 1917.*

Anybody who doesn't have fear is an idiot. It's just that you must make the fear work for you. Hell when somebody shot at me, it made me madder than hell, and all I wanted to do was shoot back.

—*Brigadier General Robin Olds, USAF.*

The most important thing in fighting was shooting, next the various tactics in coming into a fight and last of all flying ability itself.

—*Lieutenant Colonel William Avery "Billy" Bishop, RCAF.*

In nearly all cases where machines have been downed, it was during a fight which had been very short, and the successful burst of fire had occurred within the space of a minute after the beginning of actual hostilities.

—*Lieutenant Colonel William Avery "Billy" Bishop, RCAF.*

I fly close to my man, aim well and then of course he falls down.

—*Captain Oswald Boelcke, probably the world's first ace.*

Aerial gunnery is 90 percent instinct and 10 percent aim.

—*Captain Frederick C. Libby., RFC.*

When one has shot down one's first, second or third opponent, then one begins to find out how the trick is done.

—*Baron Manfred Rittmeister von Richthofen, "The Red Baron."*

You can have computer sights of anything you like, but I think you have to go to the enemy on the shortest distance and knock him down from point-blank range. You'll get him from in close. At long distance, it's questionable.

—*Colonel Erich "Bubi" Hartmann, GAF.*

Go in close, and when you think you are too close, go in closer.

—*Major Thomas B. "Tommy" McGuire, USAAF.*

I opened fire when the whole windshield was black with the enemy . . . at minimum range . . . it doesn't matter what your angle is to him or whether you are in a turn or any other maneuver.

—*Colonel Erich "Bubi" Hartmann GAF.*

As long as I look into the muzzles, nothing can happen to me. Only if he pulls lead am I in danger.

—*Captain Hans-Joachim Marseille, Luftwaffe.*

Everything in the air that is beneath me, especially if it is a one-seater . . . is lost, for it cannot shoot to the rear.

—Baron Manfred Rittmeister von Richthofen, "The Red Baron."

I started shooting when I was much too far away. That was merely a trick of mine. I did not mean so much as to hit him as to frighten him, and I succeeded in catching him. He began flying curves and this enabled me to draw near.

—Baron Manfred Rittmeister von Richthofen, "The Red Baron."

A fighter without a gun . . . is like an airplane without a wing.

—Brigadier General Robin Olds, USAF.

See, decide, attack, reverse.

—Major Erich "Bubi" Hartmann, Luftwaffe.

I was a pilot flying an airplane and it just so happened that where I was flying made what I was doing spying.

—Francis Gary Powers, U-2 reconnaissance pilot held by the Soviets for spying, in an interview after he was returned to the United States.

Fighting spirit one must have. Even if a man lacks some of the other qualifications, he can often make up for it in fighting spirit.

—Brigadier General Robin Olds, USAF.

I never went into the air thinking I would lose.

—*Commander Randy "Duke" Cunningham, USN.*

Speed is life.

—*Anon.*

Speed is the cushion of sloppiness.

—*Commander William P. "Willie" Driscoll, USNR.*

Aggressiveness was a fundamental to success in air-to-air combat and if you ever caught a fighter pilot in a defensive mood you had him licked before you started shooting.

—*Captain David McCampbell, USN, the leading U.S. Navy ace in World War II.*

The smallest amount of vanity is fatal in aeroplane fighting. Self-distrust rather is the quality to which many a pilot owes his protracted existence.

—*Captain Edward V. "Eddie" Rickenbacker.*

The experienced fighting pilot does not take unnecessary risks. His business is to shoot down enemy planes, not to get shot down. His trained hand and eye and judgment are as much a part of his armament as his machine-gun, and a fifty-fifty chance is the worst he will take—or should take—except where the show is of the kind that . . . justifies the sacrifice of plane or pilot.

—*Captain Edward V. "Eddie" Rickenbacker.*

Float like a butterfly, sting like a bee.

—*Mohammed Ali.*

A good fighter pilot, like a good boxer, should have a knockout punch . . .
You will find one attack you prefer to all others. Work on it till you can do it to
perfection . . . then use it whenever possible.

—*Captain Reade Tilley, USAAF.*

Know and use all the capabilities in your airplane. If you don't, sooner or later,
some guy who does use them all will kick your ass.

—*Dave "Preacher" Pace*

You fight like you train.

—*motto of the U.S. Navy Fighter Weapons School,* Topgun.

Fight to fly, fly to fight, fight to win.

—*motto of the U.S. Navy Fighter Weapons School,* Topgun.

The first rule of all air combat is to see the opponent first. Like the hunter
who stalks his prey and maneuvers himself unnoticed into the most favourable
position for the kill, the fighter in the opening of a dogfight must detect the
opponent as early as possible in order to attain a superior position for the attack.

—*General Adolf Galland, Luftwaffe.*

If you're in a fair fight, you didn't plan it properly.

—*Nick Lappos, chief R&D pilot, Sikorsky Aircraft.*

The British were sporting. They would accept a fight under almost all conditions.

—*Gunther Rall, Luftwaffe, 275 victories.*

A squadron commander who sits in his tent and gives orders and does not fly, though he may have the brains of Soloman, will never get the results that a man will, who, day in and day out, leads his patrols over the line and infuses into his pilots the "espirit de corps."

—*Brigadier General William "Billy" Mitchell, USAS.*

We were stripped down, even the turrets were removed. You were light and real fast, though. Our 12th squadron motto was "Alone Unarmed Unafraid." As you can imagine, this actually translated into something more like, "Alone Unarmed and Scared Shitless."

—*Theodore R. "Dick" Newell, Korean War pilot,*
12th TAC Reconnaissance Squadron, on flying the reconnaissance version of the B-26.

No guts, no glory. If you are going to shoot him down, you have to get in there and mix it up with him.

—*General Frederick C. "Boots" Blesse, USAF.*

I don't mind being called tough, because in this racket it's the tough guys who lead the survivors.

—General Curtis Emerson LeMay, USAF.

Watching the Dallas Cowboys perform, it is not difficult to believe that coach Tom Landry flew four-engine bombers during World War II. He was in B-17 Flying Fortresses out of England, they say. His cautious, conservative approach to every situation and the complexity of the plays he sends in do seem to reflect the philosophy of a pilot trained to doggedly press on according to plans laid down before takeoff. I sometimes wonder how the Cowboys would have fared all these years had Tom flown fighters in combat situations which dictated continuously changing tactics.

—Len Morgan, View From the Cockpit.

Everything I had ever learned about air fighting taught me that the man who is aggressive, who pushes a fight, is the pilot who is successful in combat and who has the best opportunity for surviving battle and coming home.

—Major Robert S. Johnson, USAAF.

The aggressive spirit, the offensive, is the chief thing everywhere in war, and the air is no exception.

—Baron Manfred Rittmeister von Richthofen, "The Red Baron."

The essence of leadership . . . was, and is, that every leader from flight commander to group commander should know and fly his airplanes.

—*Air Vice-Marshal James Edgar "Johnnie" Johnson, RAF.*

A speck of dirt on your windscreen could turn into an enemy fighter in the time it took to look round and back again. A little smear on your goggles might hide the plane that was coming in to kill you.

—*Derek Robinson,* Piece of Cake.

There are pilots and there are pilots; with the good ones, it is inborn. You can't teach it. If you are a fighter pilot, you have to be willing to take risks.

—*Brigadier General Robin Olds, USAF.*

Today it is even more important to dominate the . . . highly sophisticated weapon systems, perhaps even more important than being a good pilot; to make the best use of this system.

—*Lt. General Adolf Galland, Luftwaffe.*

An excellent weapon and luck had been on my side. To be successful, the best fighter pilot needs both.

—*Lt. General Adolf Galland, Luftwaffe.*

One of the secrets of air fighting was to see the other man first. Seeing airplanes from great distances was a question of experience and training, of knowing where to look and what to look for. Experienced pilots always saw more than the newcomers, because the latter were more concerned with flying than fighting. . . . The novice had little idea of the situation, because his brain was bewildered by the shock and ferocity of the fight.

—Air Vice-Marshal James Edgar "Johnnie" Johnson, RAF.

Only the spirit of attack borne in a brave heart will bring success to any fighter aircraft, no matter how highly developed it may be.

—Lt. General Adolf Galland, Luftwaffe.

The man who enters combat encased in solid armor plate, but lacking the essential of self-confidence, is far more exposed and naked to death than the individual who subjects himself to battle shorn of any protection but his own skill, his own belief in himself and in his wingman. Righteousness is necessary for one's peace of mind, perhaps, but it is a poor substitute for agility . . . and a resolution to meet the enemy under any conditions and against any odds.

—Major Robert S. Johnson, USAAF.

I didn't turn with the enemy pilots as a rule. I might make one turn—to see what the situation was—but not often. It was too risky.

—General John C. Meyer, Vice-Chief of Staff, USAF.

Nothing is true in tactics.

—Commander Randy "Duke" Cunningham, USN.

We were too busy fighting to worry about the business of clever tactics.

—Harold Balfour, RAF, World War I fighter pilot and British Undersecretary of State for War.

Beware the lessons of a fighter pilot who would rather fly a slide rule than kick your ass!

—Commander Ron "Mugs" McKeown, USN, Commander of the U.S. Navy Fighter Weapons School.

. . . a fighter pilot must be free to propose improvements [in tactics] or he will get himself killed.

—Commander Randy "Duke" Cunningham, USN.

Every day kill just one, rather than today five, tomorrow ten . . . that is enough for you. Then your nerves are calm and you can sleep good, you have your drink in the evening and the next morning you are fit again.

—Colonel Erich "Bubi" Hartmann, GAF.

The most important thing for a fighter pilot is to get his first victory without too much shock.

—Colonel Werner Moelders, Luftwaffe. He got his first victory and 114 others.

It is true to say that the first kill can influence the whole future career of a fighter pilot. Many to whom the first victory over the opponent has been long denied either by unfortunate circumstances or by bad luck can suffer from frustration or develop complexes they may never rid themselves of again.

—Lt. General Adolf Galland, Luftwaffe.

I gained in experience with every plane shot down, and now was able to fire in a calm, deliberate manner. Each attack was made in a precise manner. Distance and deflection were carefully judged before firing. This is not something that comes by accident; only by experience can a pilot overcome feelings of panic. A thousand missions could be flown and be of no use if the pilot has not exchanged fire with the enemy.

—Major John T. Godfrey., USAAF.

As a fighter pilot I know from my own experiences how decisive surprise and luck can be for success, which in the long run comes only to the one who combines daring with cool thinking.

—Lt. General Adolf Galland, Luftwaffe.

Months of preparation, one of those few opportunities, and the judgment of a split second are what make some pilot an ace, while others think back on what they could have done.

—Colonel Gregory "Pappy" Boyington, USMC.

Success flourishes only in perseverance—ceaseless, restless perseverance.

—Baron Manfred Rittmeister von Richthofen, "The Red Baron."

If he is superior then I would go home, for another day that is better.

—Colonel Erich "Bubi" Hartmann, GAF.

If I should come out of this war alive, I will have more luck than brains.

—Baron Manfred Rittmeister von Richthofen, "The Red Baron,"
in a letter to his mother upon being decorated with the Iron Cross.

I was struck by the joy of those pilots in committing cold-blooded murder . . . Frankly, this is not cojones. This is cowardice.

—Madeleine Albright, U.S. Ambassador to the UN, 1996.

Fighting in the air is not sport. It is scientific murder.

—Captain Edward V. "Eddie" Rickenbacker, Fighting the Flying Circus, *1919.*

I scooted for our lines, sticky with fear. I vomited brandy-and-milk and bile all over my instrument panel. Yes, it was very romantic flying, people said later, like a knight errant in the clean blue sky of personal combat.

—attributed to W.W. Windstaff, RFC, World War I.

After a scrap I usually drink my tea through a straw.

—Derek Robinson, Piece of Cake.

An Irish Airman Forsees His Death

I know that I shall meet my fate
Somewhere among the clouds above;
Those that I fight I do not hate,
Those that I guard I do not love;
My country is Kiltartan Cross,
My countrymen Kiltartan's poor,
No likely end could bring them loss
Or leave them happier than before.
Nor law, nor duty bade me fight,
Nor public men, nor cheering crowds,
A lonely impulse of delight
Drove to this tumult in the clouds;
I balanced all, brought all to mind,
The years to come seemed waste of breath,
A waste of breath the years behind
In balance with this life is death.

—William Butler Yeats.

It was no picnic despite what anyone might say later. . . . Most of us were pretty scared all the bloody time; you only felt happy when the battle was over and you were on your way home; then you were safe for a bit, anyway.

—Colin Gray, 54th Squadron RAF, World War II.

It got more exciting with each war. I mean the planes were going faster than hell when I was flying a Mustang, but by the time I got to Nam, it scared the piss out of a lot of guys just to fly the damn jets at full speed. Let alone do it in combat.

—Brigadier General Robin Olds, USAF.

Ten of My Rules for Air Fighting

- Wait until you see the whites of his eyes. Fire short bursts of 1 to 2 seconds and only when your sights are definitely "ON."
- Whilst shooting think of nothing else; brace the whole of the body; have both hands on the stick; concentrate on your ring sight.
- Always keep a sharp lookout. "Keep your finger out"!
- Height gives you the initiative.
- Always turn and face the attack.
- Make your decisions promptly. It is better to act quickly even though your tactics are not the best.
- Never fly straight and level for more than 30 seconds in the combat area.
- When diving to attack always leave a proportion of your formation above to act as top guard.
- INITIATIVE, AGGRESSION, AIR DISCIPLINE, and TEAMWORK are words that MEAN something in Air Fighting.
- Go in quickly—Punch Hard—Get out!

—Group Captain Adolphus G. "Sailor" Malan, RSAAF, World War II.

Bums on Seats

"Bums on seats" was how Captain Eddie Rickenbacker of Eastern Airlines liked to describe the airline business.

Conn McCarthy

A commercial aircraft is a vehicle capable of supporting itself aerodynamically and economically at the same time.

—*William B. Stout, designer of the Ford Tri-Motor.*

You cannot get one nickel for commercial flying.

—*Inglis M. Uppercu, founder of the first American airline to last more than a couple of months,*
Aeromarine West Indies Airways, 1923.

These days no one can make money on the goddamn airline business. The economics represent sheer hell.

—*Cyrus R. Smith, President of American Airlines.*

A recession is when you have to tighten your belt; depression is when you have no belt to tighten. When you've lost your trousers—you're in the airline business.

—*Sir Adam Thomson.*

If the Wright brother were alive today Wilber would have to fire Orville to reduce costs.

—*Herb Kelleher, Southwest Airlines,* USA Today, *June 8, 1994.*

Every other start-up wants to be another United or Delta or American. We just want to get rich.

—*Lawrence Priddy, CEO of ValueJet, 1996.*

This is a nasty, rotten business.

—Robert L. Crandall, CEO and President of American Airlines.

More than any other sphere of activity, aerospace is a test of strength between states, in which each participant deploys his technical and political forces.

—Report of the French government, 1977.

It is obvious we are fighting for the Air France Group. . . . But in actual fact, we are also fighting for France.

—Christian Blanc, Chairman of Air France, 1996.

The game we are playing here is closest to the old game of "Christians and lions."

—Robert L. Crandall.

C: I think it's dumb as hell, for Christ's sake all right, to sit here and pound the shit out of each other and neither one of us making a fucking dime.

P: *Well—*

C: I mean, goddamn! What the fuck is the point of it?

P: *Nobody asked American to serve Harlingon. Nobody asked American to serve Kansas City. . . . If you're going to overlay every route of American's on top of every route that Braniff has, I can't just sit here and allow you to bury us without giving you our best effort.*

C: Oh sure, but Eastern and Delta do the same thing in Atlanta and have for years.

P: *Do you have a suggestion for me?*

C: Yes, I have a suggestion for you. Raise your goddamn fares twenty percent. I'll raise mine the next morning. You'll make more money and I will too.

P: *Robert, we can't talk about pricing.*

C: Oh, bullshit, Howard. We can talk about any goddamn thing we want to talk about.

—Robert L. Crandall and Howard Putnam, from United States v.
American Airlines Inc. and Robert L. Crandall, U.S. District Court, CA383–0325D.

Most executives don't have the stomach for this stuff.

—Robert W. Baker, American Airlines.

Governments have supported airlines as if they were local football teams. But there are just too many of them. This is the only industry I know that has lost money consistently and makes money infrequently.

—Richard Hannah, airline analyst with UBS in London, Fortune *magazine, February 1996.*

The airline business is fast-paced, high risk, and highly leveraged. It puts a premium on things I like to do. I think I communicate well. And I am very good at detail. I love detail.

—Robert L. Crandall.

There are a lot of parallels between what we're doing and an expensive watch. It's very complex, has a lot of parts and it only has value when it's predictable and reliable.

—Gordon Bethune, Chairman and CEO of Continental Airlines, 1997.

Once you get hooked on the airline business, it's worse than dope.

—Ed Acker, while Chairman of Air Florida

We have to make you think it's an important seat—because you're in it.

—Donald Burr, founder of People Express.

I decided there must be room for another airline when I spent two days trying to get through to People Express.

—Richard Branson, founder of Virgin Atlantic.

If Richard Branson had worn a pair of steel-rimmed glasses, a double-breasted suit and shaved off his beard, I would have taken him seriously. As it was I couldn't . . .

—Lord King, Chairman of British Airways.

The trouble with predictors is that they don't know who's the prey—until he's dead.

—Sir Freddie Laker.

Freddie Laker
May be at peace with his Maker.
But he is persona non grata
With IATA.

—HRH The Duke of Edinburgh.

I really don't know one plane from the other. To me they are just marginal costs with wings.

—Alfred Kahn, airline economist and Chairman of the Civil Aeronautics Board.

There always has been a mystique and a romance about aviation, but in terms of the principles involved of satisfying your customer there's no difference between selling airline seats and chocolate bars.

—Mike "Mars Bars" Batt, British Airways
Head of Brands (Marketing) and Director of North American Routes.

If we went into the funeral business, people would stop dying.

—Martin R. Shugrue., Vice-Chairman of Pan Am.

Pan Am can go to hell.

—Alfred Kahn, airline economist and Chairman of the Civil Aeronautics Board.

You define a good flight by negatives: you didn't get hijacked, you didn't crash, you didn't throw up, you weren't late, you weren't nauseated by the food. So you're grateful.

—*Paul Theroux.*

Twenty-five per cent of the passengers of almost any aircraft show white knuckles on take-off.

—*Colin Marshall, CEO of British Airways.*

Ladies and gentleman, this is your captain speaking. We have a small problem. All four engines have stopped. We are doing our damnedest to get them going again. I trust you are not in too much distress.

—*Captain Eric Moody, British Airways, after flying through volcanic ash in a B-747.*

The British Islands are small islands and our people numerically a little people. Their only claim to world importance depends upon their courage and enter-prise, and a people who will not stand up to the necessity of air service planned on a world scale, and taking over thousands of aeroplanes and thousands of men from the onset of peace, has no business to pretend anything more than a second rate position in the world. We cannot be both Imperial and mean.

—*H. G. Wells, minority report of the committee to study "the development and regulation after the war of aviation for civil and commercial purposes from a domestic, an imperial and an international stand-point," 1917.*

It may be questioned whether civil aviation in England is to be regarded as one of those industries which is unable to stand on its own two feet, and is yet so essential to the national welfare that it must be kept alive at all costs.

—Major-General Sir Frederick Sykes, first Director of British Civil Aviation, 1919.

. . . It wasn't until the jet engine came into being and that engine was coupled with special airplane designs—such as the swept wing—that airplanes finally achieved a high enough work capability, efficiency and comfort level to allow air transportation to really take off.

—Joseph F. Sutter, Boeing Commercial Airplanes.

We're going to make the best impression on the traveling public, and we're going to make a pile of extra dough just from being first.

—Cyrus R. Smith, on the introduction of the Boeing 707, Forbes, *1956.*

There has always been a certain romanticism associated with the airline business. We must avoid its perpetuation at Eastern at all costs.

—Frank Borman, Eastern Airlines.

We are long-term players in the industry. We're not just crazy and emotional. We try to be logical business managers.

—Frank Lorenzo, Eastern Airlines.

I'm not paid to be a candy ass. I'm paid to go and get a job done. I could have ended up with another job, but the job I ended up with was piecing together a bunch of companies that were all headed for the junk heap . . . . I've got to be the bastard who sits around Eastern Airlines and says, "Hey, we're losing $3 million a day or whatever the number is and bang, bang, bang, bang, what do you do?" So, some jobs are easier than others.

—*Frank Lorenzo, Eastern Airlines.*

I can't imagine a set of circumstances that would produce Chapter 11 for Eastern.

—*Frank Lorenzo, Eastern Airlines.*

As a businessman, Frank Lorenzo gives capitalism a bad name.

—*William F. Buckley.*

From this day forward, you must assume that Eastern Airlines intends to force a strike, and you must be prepared for the worst.

—*Charlie Bryan, Eastern Airlines machinists union leader.*

We were raped!

—*Frank Borman, after capitulating to Charlie Bryan's wage demands.*

Frank is capable of any kind of behavior to win.

—*Don Burr.*

Regulation has gone astray. . . . Either because they have become captives of regulated industries or captains of outmoded administrative agencies, regulators all too often encourage or approve unreasonably high prices, inadequate service, and anticompetitive behavior. The cost of this regulation is always passed on to the consumer. And that cost is astronomical.

—Senator Edward Kennedy, opening remarks to the
Subcommittee on Administrative Practice and Procedure, February 6, 1975.

Whenever competition is feasible it is, for all its imperfections, superior to regulation as a means of serving the public interest.

—Alfred Kahn, airline economist and Chairman of the Civil Aeronautics Board.

British Airways believes that it is intrinsically deceptive for two carriers to share a designator code.

—Comments of British Airways on PDSR-85, Notice of Proposed Rulemaking, Docket 42199, 1984.

I've tried to make the men around me feel, as I do, that we are embarked as pioneers upon a new science and industry in which our problems are so new and unusual that it behoves no one to dismiss any novel idea with the statement that "it can't be done!" Our job is to keep everlasting at research and experimentation, to adapt our laboratories to production as soon as possible, and to let no new improvement in flying and flying equipment pass us by.

—William E. Boeing, founder The Boeing Company, 1929.

A sick customer results in a sick airplane manufacturing industry, whatever the cause may be.

—*John E. Steiner, Boeing Commercial Airplanes.*

We built a jet airplane to get in and out of a 5,000-ft field. No one believed it could be done.

—*Joseph F. Sutter, Boeing Commercial Airplanes, on the B-727.*

This is the most important aviation development since Lindbergh's flight. In one fell swoop, we have shrunken the earth.

—*Juan Trippe, founder of Pan Am, on the introduction of jet aircraft.*

If the pilots were in charge, Columbus would still be in port. They believe the assertion that the world is flat.

—*Robert L. Crandall, American Airlines, 1993.*

I've said many times that I'd be thrilled to sell the airline to the employees and our guys said no, we'll take all the money, anyway.

—*Robert L. Crandall, American Airlines, 1997.*

The Wright Amendment is a pain in the ass, but not every pain in the ass is a constitutional infringement.

—*Herb Kelleher, Southwest Airlines.*

Think and act big and grow smaller, or think and act small and grow bigger.

—Herb Kelleher, Southwest Airlines.

It takes nerves of steel to stay neurotic.

—Herb Kelleher, Southwest Airlines.

You [the employees] are involved in a crusade.

—Herb Kelleher, Southwest Airlines.

That place [Southwest] runs on Herb Kelleher's bullshit.

—Robert W. Baker, American Airlines.

Be Luke Skywalker, not Darth Vader. Ultimately love is stronger than evil.

—Donald Burr, founder of People Express.

Those were my children being slaughtered.

—Donald Burr, founder of People Express.

The state of our airline industry is a national embarrassment.

—Tom Plaskett, Chairman of Pan Am, following the airline's collapse.

I think historically, the airline business has not been run as a real business. That is, a business designed to achieve a return on capital that is then passed on to shareholders. It has historically been run as an extremely elaborate version of a model railroad, that is, one in which you try to make enough money to buy more equipment.

—Michael Levine, Executive Vice President of Northwest Airlines, 1996.

It's not a testosterone-driven industry any longer. Success is making money, not in the size of the airline.

—Gordon Bethune, Chairman and CEO of Continental Airlines, 1996.

In a sense, when we started Virgin Atlantic, I was trying to create an airline for myself. If you try to build the perfect airline for yourself, it will be appreciated by others.

—Richard Branson, 1996.

Branson's "dirty tricks" claim unfounded.

—Headline of an article in the British Airways newsletter BA News, *1992.*
The article became the center of the largest libel payment in British legal history.

Sue the bastards.

—Sir Freddie Laker's advice to Richard Branson regarding
British Airways' dirty tricks campaign against Virgin Atlantic.

It was over in a blink of an eye, that moment when aviation stirred the modern imagination. Aviation was transformed from recklessness to routine in Lindbergh's lifetime. Today the riskiest part of air travel is the drive to the airport, and the airlines use a barrage of stimuli to protect passengers from ennui.

—*George Will,* Charles Lindbergh, Craftsman, *May 15, 1977.*

The air is annoyingly potted with a multitude of minor vertical disturbances which sicken the passengers and keep us captives of our seat belts. We sweat in the cockpit, though much of the time we fly with the side windows open. The airplanes smell of hot oil and simmering aluminum, disinfectant, feces, leather, and puke . . . the stewardesses, short-tempered and reeking of vomit, come forward as often as they can for what is a breath of comparatively fresh air.

—*Ernest K. Gann, describing airline flying in the 1930s.*

Fly the friendly skies.

—*United Airlines advertising slogan.*

Something special in the air.

—*American Airlines advertising slogan.*

To fly. To serve.

—*British Airways advertising slogan.*

The world's favourite airline.

—British Airways advertising slogan.

We love to fly. And it shows.

—Delta Air Lines advertising slogan.

Some people just know how to fly.

—Northwest Airlines advertising slogan.

How do we love you? Let us count the ways . . .

—Early Southwest Airlines advertising slogan.

The Proud Bird with the Golden Tail

—Advertising slogan of the "real" Continental Airlines, pre-1983 bankruptcy.

The Wings of Man.

—Eastern Airlines advertising slogan.

If God wanted us to fly, He would have given us tickets.

—Mel Brooks.

If God had really intended men to fly, He'd make it easier to get to the airport.

—George Winters.

In the space age, man will be able to go around the world in two hours—one hour for flying and one hour to get to the airport.

—Neil McElroy.

In America there are two classes of travel—first class, and with children.

—Robert Benchley.

United hired gentlemen with the expectation of training them to become pilots, Northwest hired pilots hoping to train them to become gentlemen. To date, despite their best efforts, neither carrier can be considered successful.

—Edward Thompson.

Life expands in an aeroplane. The traveler is a mere slave in a train, and, should he manage to escape from this particular yoke, the car and the ship present him with only limited horizons. Air travel, on the other hand, makes it possible for him to enjoy the "solitary delights of infinite space." The earth speeds below him, with nothing hidden, yet full of surprises. Introduce yourself to your pilot. He is always a man of the world as well as a flying ace.

—Early French advertisement for airline service.

Clichés

Aviate, Navigate, Communicate.

Truly superior pilots are those who use their superior judgment to avoid those situations where they might have to use their superior skills.

Rule one: No matter what else happens, fly the airplane.

Flying is hours of boredom, punctuated by moments of stark terror.

Fly it until the last piece stops moving.

It's better to be down here wishing you were up there, than up there wishing you were down here.

An airplane will probably fly a little bit overgross but it sure won't fly without fuel.

Believe your instruments.

Think ahead of your airplane.

I'd rather be lucky than good.

The propeller is just a big fan in the front of the plane to keep the pilot cool. Want proof? Make it stop; then watch the pilot break out into a sweat.

If we are what we eat, then some pilots should eat more chicken.

If you're ever faced with a forced landing at night, turn on the landing lights to see the landing area. If you don't like what you see, turn 'em back off.

A check ride ought to be like a skirt, short enough to be interesting but still be long enough to cover everything.

Standard checklist philosophy requires that pilots read to each other the actions they perform every flight, and recite from memory those they need every 3 years.

Experience is the knowledge that enables you to recognize a mistake when you make it again.

There are some flight instructors where the student is important, and there are some instructors where the instructor is important. Pick carefully.

Speed is life; altitude is life insurance.

No one has ever collided with the sky.

Lack of planning on your part does not constitute an emergency on mine.

There's no place like cloudbase

One peek is worth a thousand instrument cross-checks.

Always remember you fly an airplane with your head, not your hands.

Never let an airplane take you somewhere you brain didn't get to 5 minutes earlier.

If it's red or dusty don't touch it.

Don't drop the aircraft in order to fly the microphone.

An airplane flies because of a principle discovered by Bernoulli, not Marconi.

Cessna pilots are always found in the wreckage with their hand around the microphone.

If you push the stick forward, the houses get bigger; if you pull the stick back, they get smaller.

To go up, pull the stick back. To go down, pull the stick back harder.

Hovering is for pilots who love to fly but have no place to go.

Flying is the second greatest thrill known to man. . . . Landing is the first!

Every one already knows the definition of a "good" landing is one from which you can walk away. But very few know the definition of a "great landing." It's one after which you can use the airplane another time.

The probability of survival is equal to the angle of arrival.

There are two types of tailwheel (or retractable gear) pilot: those who have ground-looped (or landed gear up) and those that will.

If you've got time to spare, go by air.
(More time yet? Go by jet.)

IFR: I Follow Roads.

There are old pilots, and there are bold pilots, but there are no old, bold pilots.

You know you've landed with the wheels up when it takes full power to taxi.

If you don't gear up your brain before takeoff, you'll probably gear up your airplane on landing.

Navy carrier pilots regards Air Force pilots: "Flare to land, squat to pee."

Air Force pilots regards Navy carrier pilots: "Next time a war is decided by how well you land on a carrier, I'm sure our Navy will clean up. Until then, I'll worry about who spends their training time flying and fighting."

A kill is a kill.

He who sees first lives longest.

Fighter pilots make movies; attack pilots make history.

In thrust I trust.

Jet noise: The sound of freedom.

I had a fighter pilot's breakfast—two aspirin, a cup of coffee, and a puke.

Those who hoot with the owls by night should not fly with the eagles by day.

Fly with the eagles, or scratch with the chickens.

It only takes two things to fly: airspeed and money.

Forget all that stuff about thrust and drag, lift, and gravity. An airplane flies because of money.

Do you see that propeller? Well, everything behind it revolves around money.

A smooth touchdown in a simulator is as exciting as kissing your sister.

Experience is a hard teacher. First comes the test, then the lesson.

A helicopter is a collection of rotating parts going round and round and reciprocating parts going up and down—all of them trying to become random in motion.

Helicopters can't really fly—they're just so ugly that the earth immediately repels them.

Helicopters don't fly. They beat the air into submission.

Chopper pilots get it up quicker.

Helicopters don't fly, they just vibrate against the earth, and the earth rejects them into the air.

Never fly in anything that has the wings traveling faster than the fuselage. Like a helicopter.

The owner's guide that comes with a $500 refrigerator makes more sense than the one that comes with a $50 million airliner.

If it doesn't work, rename it. If that doesn't help, the new name isn't long enough.

Federal Aviation Regulations are worded either by the most stupid lawyers in Washington, or the most brilliant.

The only thing that scares me about flying is the drive to the airport.

Flying is not Nintendo. You don't push a button and start over.

Life is lead points and habit patterns.

Gravity: killer of young adults.

I'm not speeding officer—I'm just flying low.

Young man, was that a landing or were we shot down?

Sorry, folks, for the hard landing. It wasn't the pilot's fault, and it wasn't the plane's fault. It was the asphalt.

Learn from the mistakes of others. You won't live long enough to make all of them yourself.

Three things kill young pilots in Alaska—weather, weather, and weather.

Please don't tell Mum I'm a pilot; she thinks I play piano in a whorehouse.

Pilots believe in clean living. They never drink whiskey from a dirty glass.

FAA regulations forbid drinking within 8 feet of the aircraft and smoking within 50 hours of flight. Or is it the other way around?

"Please see me at once" memos from the chief pilot are distributed on Fridays after office hours.

Fly low and slow and don't tip on the turns.

An accident investigation hearing is conducted by nonflying experts who need 6 months to itemize all the mistakes made by a crew in the 6 minutes it had to do anything.

Things Which Do You No Good in Aviation

Altitude above you.

Runway behind you.

Fuel in the truck.

A navigator.

Half a second ago.

Approach plates in the car.

The airspeed you don't have.

The more traffic at an airport, the better it is handled.

If man were meant to fly, God would have given him baggy, Nomex skin.

If God meant man to fly, He'd have given him more money.

What's the difference between God and pilots? God doesn't think he's a pilot.

Will Rogers never met a fighter pilot.

To err is human, to forgive is divine; neither of which is Air Force policy.

Flying is not dangerous; crashing is dangerous.

You can land anywhere once.

Flying is the perfect vocation for a man who wants to feel like a boy, but not for one who still is.

There are four ways to fly: the right way, the wrong way, the company way and the captain's way. Only one counts.

A good simulator check ride is like successful surgery on a cadaver.

Asking what a pilot thinks about the FAA is like asking a fireplug what it thinks about dogs.

Crime wouldn't pay if the FAA took it over and would go bankrupt if an airline management did.

Trust your captain . . . but keep your seat belt securely fastened.

An airplane may disappoint a good pilot, but it won't surprise him.

Winds-aloft reports are of incomparable value—to historians.

Any pilot who relies on a terminal forecast can be sold the Brooklyn (or London) Bridge. If he relies on winds-aloft reports, he can be sold Niagara Falls (or The Tower of London).

The difference between flight attendants and jet engines is that the engine usually quits whining when it gets to the gate.

The friendliest stewardesses are those on the trip home.

Out on the line, all the girls are looking for husbands and all the husbands are looking for girls.

The most nerve-wracking of airline duties: the flight engineer's job on a proving run flown by two chief pilots.

Good judgment comes from experience, and experience comes from bad judgment.

Being an airline pilot would be great if you didn't have to go on all those trips.

Aviation is not so much a profession as it is a disease.

The nicer an airplane looks, the better it flies.

Remember when sex was safe and flying was dangerous?

Renting airplanes is like renting sex: It's difficult to arrange on short notice on Saturday, the fun things always cost more, and someone's always looking at their watch.

Jet and piston engines work on the same principle: suck, squeeze, bang, blow.

You can always depend on twin-engine aircraft. When the first engine quits the second will surely fly you to the scene of an accident.

The real value of twin engine aircraft is it will double your chances of engine failure.

If it ain't broke, don't fix it; if it ain't fixed, don't fly it.

A mechanic's favorite: That's not a leak; its a seep.

About aerobatics: It's like having sex and being in a car wreck at the same time.

If it's ugly, it's British; if it's weird, it's French; and if it's ugly and weird, it's Russian.

A male pilot is a confused soul who talks about women when he's flying and about flying when he's with a woman.

A grease-job landing is 50 percent luck; two in a row are entirely luck; three in a row and someone's lying.

There are three simple rules for making a smooth landing: Unfortunately, no one knows what they are.

It's a good landing if you can still get the doors open.

First, listen to the question the student asked, then listen to the question he didn't ask and then figure out the question he really meant to ask.

Airspeed, altitude, or brains; you always need at least two.

A groundschool instructor understands piloting the way an astronomer understands the stars.

Every groundschool class includes one ass who, at 5 minutes before 5, asks a question requiring a 20-minute explanation.

Gravity: it's not just a good idea; it's the law.

The law of gravity is not a general rule.

You can only tie the record for flying low.

Flying at night is the same as flying in the day, except you can't see.

It is easier to cope with a single in-flight problem than a series of minor ones. Real trouble must be swallowed in small doses.

It's no wonder England serves beer warm, Lucas manufactures most of their refrigeration equipment.

If at first you don't succeed, well, so much for skydiving.

I want to die like my grandfather did, peacefully in his sleep. Not screaming in terror like his passengers.

I give that landing a 9 . . . on the Richter scale.

It's better to break ground and head into the wind than to break wind and head into the ground.

Modern air travel would be very enjoyable . . . if I could only learn to enjoy boredom, discomfort and fatigue.

Passengers prefer old captains and young stewardesses.

A captain with little confidence in his crew usually has little in himself.

The only soul more pitiful than a captain who cannot make up his mind is the copilot who has to fly with him.

The sharpest captains are the easiest to work with.

The only thing worse than a captain who never flew as copilot is a copilot who once was a captain.

Be nice to your first officer. He may be your captain at your next airline.

A copilot is a knothead until he spots opposite direction traffic at 12 o'clock, after which he's a goof-off for not seeing it sooner.

A captain is two flight engineers sewn together.

Everything in the company manual—policy, warnings, instructions, the works—can be summed up to read, "Captain, it's your baby."

Nothing is more optimistic than a dispatcher's estimated time of departure.

Clocks lie; an 18-hour layover passes much quicker that an 8-hour day.

The worst day of flying still beats the best day of real work.

Any pilot who does not privately consider himself the best in the game is in the wrong game.

It's best to keep the pointed end going forward as much as possible.

Assumption is the mother of all fuck-ups.

Gravity is bullshit: The Earth sucks.

It's better to die than to look bad, but it is possible to do both.

Death is a small price to pay for looking shit hot.

If God had intended man to fly, he would have given him enough money for a Bonanza.

What do you call a pregnant flight attendant? Pilot error.

What separates flight attendants from the lowest form of life on earth? The cockpit door.

The three most common phrases in airline aviation are: "Was that for us?" "What'd he say?" and "Oh Shit!" Since computers are now involved in flying, a new one has been added: "What's it doing now?"

The most sensitive mechanism in modern aviation is the shower control in a layover hotel.

Any attempt to stretch fuel is guaranteed to increase headwinds.

Any comment about how well things are going is an absolute guarantee of trouble.

A terminal forecast is a horoscope with numbers.

A thunderstorm is never as bad on the inside as it appears on the outside. It's worse.

Below 20, boys are too rash for flying; above 25, they are too prudent.

Son, I was flying airplanes for a living when you were still in liquid form.

Eagles may soar, but weasels never get sucked into jet air intakes.

Most airline food tastes like warmed-over chicken because that's what it is.

Everything is accomplished through teamwork until something goes wrong; then one pilot gets all the blame.

Let's make a 360 and get the hell out of here!?!

Don't trust nobody and don't do nothing dumb.

One who flies with fear encourages fate.

It's easy to make a small fortune in aviation. You start with a large fortune.

Pilots are just plane people with a special air about them.

There I was at 40,000 ft. when the autopilot jumped out with the only parachute on board and left me with nothing but a silkworm and a sewing kit.

There I was at 15,000 ft. with nothing on the clock but the maker's name—and that was on the back and peeling.

There I was; fog was so thick I couldn't see the instruments. Only way I knew I was inverted was my flying medals were in my eyes. But I knew I was really in trouble when the tower called me and told me to climb and maintain field elevation.

When the last Blackhawk helicopter goes to the boneyard, it will be on a sling under a Huey.

Speed is money in your pocket, altitude is money in the bank.

Keep the shiny side up and the greasy side down.

Don't forget to keep the blue side up.

A fool and his money are soon flying more airplane than he can handle.

Definition of a complex airplane: landing a taildragger on pavement with a 20-kn quartering crosswind.

When a forecaster talks about yesterday's weather, he's a historian; when he talks about tomorrow's, he's reading tea leaves.

The main thing is to take care of the main thing.

A thunderstorm is nature's way of saying, "Up yours."

Learning a little about flying is like leading a tiger by the tail—the end does not justify the means.

The last thing every pilot does before leaving the aircraft after making a gear up landing is to put the gear selection lever in the "down" position.

Remember, you're always a student in an airplane.

Keep looking around; there's always something you've missed.

Fuel in the tanks is limited. Gravity is forever.

Try to keep the number of your landings equal to the number of your takeoffs.

Takeoffs are optional. Landings are mandatory.

Son, if you're trying to impress me with your flying, relax. Most of the time I can't even impress myself.

Flight Instructor Favorites

You don't know what you don't know.

Much of what you think you know is incorrect.

Together, we must find out why you don't know what you don't know.

It is practice of the right kind that makes perfect.

You will never do well if you stop doing better.

Students never fail; only teachers do.

A student's performance is not so much a reflection on the student as it is on the instructor's ability to teach.

Learning is not a straight line up . . . let the teacher set the standards of performance.

Much of learning to fly is to unlearn preconceptions and habits.

The way you are first taught and learn a procedure is the way you will react in an emergency. It's important to learn right the first time.

Unlearning is a very necessary and difficult part of learning to fly.

You learn according to what you bring into the situation.

Being prepared for a flight saves you money by saving time.

Given the choice, make the safe decision.

If you must make a mistake, make it a new one.

One problem is a problem; two problems are a hazard; three problems create accidents.

Trusting to luck alone is not conducive to an extended flying career.

We progress through repeated success; we learn through our mistakes.

An instructors knowledge is proportional to the mistakes he's made.

Good habits deteriorate over time.

Accidents happen when you run out of experience.

Self-instruction is the garden that raises bad habits.

Our failures teach us. If you want to increase your chances of success double your failure rate.

. . . almost always. Nothing is always.

Luck will do for skill, but not consistently.

The nice thing about a mistake is the pleasure it gives others.

You're only young once, but you can be immature forever.

Flying, like life, is full of precluded possibilities.

Can't do . . . won't do . . . shouldn't do . . .

What you know is not as important as what you do with it.

.......Food for Thought

In days gone by, I've proved my worth
By zooming low across the earth.
I've buzzed the valleys and the mountain ridges,
I've dove my craft beneath the bridges.
I've looped and spun and rolled my wings,
I've sung the songs that pilots sing.
I've tried most stunts, it must be said,
Yet never learnt to use my head.
So here's a toast—To you and me!
But you drink both, I'm dead . . . you see.

—Anon.

Bite into my wing and don't say anything but "2," "bingo," and "Lead, you're on fire."

—Briefing to a novice USAF wingman: stay close, acknowledge channel changes, tell me when you're out of gas, and let me know if there is something wrong with my aircraft. Otherwise, shut up.

Son, your wife's legs have more time in the air than you do.

—Welcome to a new copilot from an old captain.

Son, I've got more time sitting on the lav in this airliner than you have total time.

—*Welcome to a new copilot from an old captain.*
Also heard as, "I've got more time in the flare . . . " and "I've got more time in the bunk . . ."

Throttle back, son. You're not going to make the boat go any faster.

—*Air bosses on aircraft carriers to flight students on initial carrier qualifications*
who stay at maximum power after they have been jerked to a stop by the arresting gear.

You've got to land here, son. This is where the food is.

—*Unknown landing signal officer to carrier pilot after his 6th unsuccessful landing.*

I ran out of altitude, airspeed, and ideas all at the same time.

—*When asked why he ejected. Attributed to Tony Lavier,*
Chuck Yeager, and just about every other well-known test pilot.

Miscellaneous

Conn McCarthy

It is not necessarily impossible for human beings to fly, but it so happens that God did not give them the knowledge of how to do it. It follows, therefore, that anyone who claims that he can fly must have sought the aid of the devil. To attempt to fly is therefore sinful.

—Roger Bacon, thirteenth-century Franciscan friar.

Words are heavy like rocks . . . they weigh you down. If birds could talk, they wouldn't be able to fly.

—Marilyn, from the TV show Northern Exposure.

Any language where the unassuming word *fly* signifies an annoying insect, a means of travel, and a critical part of a gentleman's apparel is clearly asking to be mangled.

—Bill Bryson, first page of Chapter 1, Mother Tongue: The English Language, *1990.*

The English, a haughty nation, arrogate to themselves the empire of the sea; the French, a buoyant nation, make themselves masters of the air.

—The Count of Provence (afterward Louis XVIII of France), Impromptu on the first successful balloon ascension by the brothers Montgolfier, 1783. French Original, "Les Anglais, nation trop fière, S'arrogent l'empire des mers; Les Français, nation légère, S'emparent de celui des airs."

Providence has given to the French the empire of the land, to the English that of the sea, and to the Germans that of the air.

—Jean Paul Richter, quoted by Thomas Carlyle, Edinburgh Review, *1827.*

Not the cry, but the flight, of the wild duck leads the flock to fly and follow.

—Chinese proverb.

The miracle is not to fly in the air, or to walk on the water, but to walk on the earth.

—Chinese proverb.

The airplane stays up because it doesn't have the time to fall.

—Orville Wright.

If you don't get in that plane you'll regret it. Maybe not today, maybe not tomorrow, but soon and for the rest of your life.

—Rick Blaine, in the movie Casablanca, *1942.*

As a piece of applied science the aeroplane has a place alongside the wheel, gunpowder, the printing press and the steam engine as one of the great levers of change in world history. The effect of aircraft on the way we live has been profound: they have shrunk the world, mingling previously isolated cultures, they have added a menacing dimension to warfare, spawned new technologies, created new economic zones and given us a toehold in Space.

—Iven Rendall, first paragraph of the Introduction, Reaching for the Skies, *1988.*

Aviation is for grown men, alert, strong and above all capable of endurance.

—Charles Turner, holder of many early aviation records.

Facts are the air of scientists. Without them you can never fly.

—Ivan Pavlov.

Terminals have always been, and probably always will be the "bottle-necks" of transportation, whether of ground, water, or air systems.

—Harry H. Blee, U.S. Aeronautics Branch, 1932.

The airport runway is the most important mainstreet in any town.

—Norm Crabtree, aviation director for the state of Ohio.

There is no other airport in the world which serves so many people and so many airplanes. This is an extraordinary airport . . . it could be classed as one of the wonders of the modern world.

—President John F. Kennedy, dedicating Chicago's O'Hare Airport, March 23, 1963.

There's nothing like an airport for bringing you down to earth.

—Richard Gordon.

It's no coincidence that in no known language does the phrase "As pretty as an airport" appear.

—Douglas Adams, The Long Dark Tea-Time of the Soul.

[London's] Heathrow has been described as the only building site to have its own airport.

—*Anon.*

. . . Hell, which as every frequent traveler knows, is in Concourse D of O'Hare Airport.

—*Dave Barry. There is no Concourse D at O'Hare.*

Airplane travel is nature's way of making you look like your passport photo.

—*Vice President Albert Gore.*

Flight was a metaphor for the new Nietzschean age that was dawning. The deeds of the technological heroes of the twentieth century would equal, and perhaps exceed, those of the mythical figures of the Ancient World. . . . The urge to dominate, to master, to conquer, was the motivation that drove men to fly. Speed was the divinity of the new century, to be worshipped at any cost. The cult of movement required victims. In its service, no sacrifice was too great. Aviators were the new aristocracy. Power and primacy would come to those peoples who dominated the air. . . . Death was the price that man would have to pay in order to live like gods in a world of fast machines.

—*Robert Wohl, last paragraph of* A Passion For Wings:
Aviation and the Western Imagination 1908–1918, *1994.*

I always thought that my airplane conveyed a silent sermon. To the earthbound observer, its silhouette was the shape of the cross on which Jesus was crucified.

—*E. R. Trimble.*

When wild the head-wind beat,
Thy sovereign Will commanding
Bring them who dare to fly
To a safe landing.

—*Duncan Campbell Scott, RCAF,* Hymn for Those in the Air.

No, son—you're not up there alone—not with all the things you come through. You have the greatest copilot in the world even if there is just room for one in that fighter ship—no, you're not alone.

—*Colonel Robert L. Scott, Jr., USAAF,* God Is My Copilot.

We who fly are going to get to know that Great Flying Boss in the sky better and better.

—*Colonel Robert L. Scott, Jr., USAAF,* God Is My Copilot.

The bulk of mankind is as well equipped for flying as thinking.

—*Jonathan Swift.*

Which is now a more hopeful statement that Swift intended it to be.

—*Will Durant.*

That's not flying; that's just falling with style.

—Woody, in the movie Toy Story, *with regard to Buzz Lightyear, 1996.*

If Beethoven had been killed in a plane crash at the age of 22, it would have changed the history of music . . . and of aviation.

—Tom Stoppard.

Arguing with a pilot is like wrestling with a pig in the mud; after a while you begin to think the pig likes it.

—Seen on a General Dynamics' bulletin board.

There are many excellent pilots who would rather do anything than land a private airplane at Newark, Cleveland, or Chicago.

*—*Aviation *magazine, August 1935.*

The vilest enemy of the morale of aeronautics is a scab.

—David Behnecke, founder of the Air Line Pilots Association.

If a prisoner, he must escape; if dead, he must come back to life.

*—*Le Temps *newspaper referring to the disappearance of French World War I ace Capitaine Georges Guynemer.*

The ships hung in the sky in much the same way that bricks don't.

—Douglas Adams, The Hitchhiker's Guide to the Galaxy.

I am not afraid of crashing; my secret is . . . just before we hit the ground, I jump as high as I can.

—*Bill Cosby.*

Cocooned in Time, at this inhuman height,
The packaged food tastes neutrally of clay.
We never seem to catch the running day
But travel on in everlasting night . . .

—*John Betjeman.*

If helicopters are so safe, how come there are no vintage/classic helicopter fly-ins?

—*Jim Tavenner.*

Lady, you want me to answer you if this old airplane is safe to fly? Just how in the world do you think it got to be this old?

—*Jim Tavenner.*

The scientific theory I like best is that the rings of Saturn are composed entirely of lost airline luggage.

—*Mark Russell.*

You know they invented wheelbarrows to teach FAA inspectors to walk on their hind legs.

—*Marty Caidin.*

You cannot fly like an eagle with the wings of a wren.

—*William Henry Hudson,* Afoot in England, *1909.*

There is no more alluring airspace in the world than the slit up a China girl's dress.

—*Ernest K. Gann,* Band of Brothers.

If you want to fuck with the eagles, you have to learn to fly.

—*Veronica, in the movie* Heathers, *1989.*

The pilot's life is founded on three things: sex, seniority, and salary, in that order.

—*Dr. Ludwig Lederer, corporate physician at American Airlines.*

We have no effective screening methods to make sure pilots are sane.

—*Dr. Herbert Haynes, Federal Aviation Authority.*

The wish to be able to fly is to be understood as nothing else than a longing to be capable of sexual performance.

—*Sigmund Freud.*

A free ride and free food are two of the three things no pilot ever turns down.

—*Attributed to Dick Rutan.*

Why is Communism like flying in an aeroplane?
You see the glorious horizon approaching, but the longer you fly the less the glorious horizon seems to approach, you feel sick, and you can't get out.

—Anon.

When asked "How was your flight?'
Well, aeronautically it was a great success. Socially it left quite a bit to be desired.

—Noel Coward.

Airplanes are like women—pick what you like and try to get it away from the guy who has it, then dress it out to the limit of your wallet and taste.

—Stephen Coonts, The Cannibal Queen.

My first wife didn't like to fly, either.

—Gordon Baxter.

Flying around the world is like raising kids. When you've finally figured out how to do it the right way, you've finished.

—Ron Bower, who has flown around the world solo in a helicopter.

The mother eagle teachers her little ones to fly by making their nest so uncomfortable that they are forced to leave it and commit themselves to the unknown world of air outside. And just so does our God to us.

—Hannah Whitall Smith.

Hold fast to dreams, for if dreams die life is a broken winged bird that cannot fly.

—*Langston Hughes.*

In spite of me it drew forward into the wind, notwithstanding my resistance it tended to rise. Thus I have discovered the secret of the bird and I comprehend the whole mystery of flying.

—*Jean Marie Le Bris, a French sea captain who experimented with gliders, circa 1850.*

This is earth again, the earth where I've lived and now will live once more . . . I've been to eternity and back. I know how the dead would feel to live again.

—*Charles A. Lindbergh, on sighting Ireland after his first solo Atlantic crossing, 1927.*

Of all the inventions that have helped to unify China perhaps the airplane is the most outstanding. Its ability to annihilate distance has been in direct proportion to its achievements in assisting to annihilate suspicion and misunderstanding among provincial officials far removed from one another or from the officials at the seat of government.

—*Madame Chiang Kai-shek, "Wings Over China,"* Shanghai Evening Post, *March 12, 1937.*

I have seen so much on my pilgrimage through my three score years and ten,
That I wouldn't be surprised to see a railroad in the air
Or a Yankee in a flyin' ship a-goin' most anywhere.

—*J. H. Yates,* The Old Ways and the New.

On wings of winds came flying all abroad.

—*Alexander Pope,* Epistle to Dr. Arbuthnot.

Flying a plane is no different from riding a bicycle. It's just a lot harder to put baseball cards in the spokes.

—*Captain Rex Kramer, in the movie* Airplane!, *1980.*

We have clearance, Clarence. Roger, Roger. What's our vector, Victor?

—*Cockpit crew, in the movie* Airplane!, *1980.*

The odds against there being a bomb on a plane are a million to one, and against two bombs a million times a million to one. Next time you fly, cut the odds and take a bomb.

—*Benny Hill.*

Let brisker youths their active nerves prepare
Fit their light silken wings and skim the buxom air.

—*Richard Owen Cambridge,* Scriblerad, *1751.*

To propel a dirigible balloon through the air is like pushing a candle through a brick wall.

—*Alberto Santos-Dumont, with regard to Zeppelin's airship.*

The Boeing 747 is so big that it has been said that it does not fly; the earth merely drops out from under it.

—Captain Ned Wilson, Pan Am.

If they could get a washing machine to fly, my Jimmy could land it.

—Blanche Lovell, in the movie Apollo 13, 1995.

Military pilots and then, soon, airline pilots, pilots from Maine and Massachusetts and the Dakotas and Oregon and everywhere else, began to talk in that poker-hollow West Virginia drawl, or as close to it as they could bend their native accents. It was the drawl of the most righteous of all the possessors of the right stuff: Chuck Yeager.

—Tom Wolfe, The Right Stuff.

Not so long ago, when I was a student in college, just flying an airplane seemed a dream. But that dream turned into reality.

—Charles A. Lindbergh, beginning of his autobiography, The Spirit of St. Louis, 1953.

The work of the individual still remains the spark that moves mankind ahead.

—Igor I. Sikorsky.

Any damned fool can criticize, but it takes a genius to design it in the first place.

—Edgar Schmued, chief designer at North American Aviation.

Some newspapers have an adversarial approach to The Boeing Company that actually nauseates me and I've stopped reading them. I spent fifteen years on the Boeing crash investigation committee, and I learned first hand the difference between what gets reported in the paper and what the facts are. I concluded that there was almost no relationship between what was written there and the facts, and it kind of made me nervous about reading anything else. I just quit taking the papers.

—Granville "Granny" Frazier, The Boeing Company.

Some are concerned about the risks from computer hackers with such a connected system. [A spokesman] said that with the current FAA software, it's not a problem. A recent White House panel on security concluded that [the] software is so out of date that no one could possibly hack into it.

—Aviation Week & Space Technology, *December 1996.*

They're multipurpose. Not only do they put the clips on, but they take them off.

—Pratt & Whitney spokesperson explaining why the company charged the Air Force nearly $1000 for an ordinary pair of pliers, 1996.

I don't like flying because I'm afraid of crashing into a large mountain. I don't think Dramamine is going to help.

—Kaffie, in the movie A Few Good Men, *1992.*

Angels can fly because they take themselves lightly.

—G. K. Chesterton, Orthodoxy, *1908.*

The important thing in aeroplanes is that they shall be speedy.

—Baron Manfred Rittmeister von Richthofen, "The Red Baron."

An aircraft which is used by wealthy people on their expense accounts, whose fares are subsidized by much poorer taxpayers.

—Denis Healey, British Labour Party, on the Concorde.

We realized the difficulties of flying in so high a wind, but estimated that the added dangers in flight would be partly compensated for by the slower speed in landing.

—The Wright Brothers.

I found myself caught in them wires and the machine blowing across the beach heading for the ocean, landing first on one end and then on the other, rolling over and over, and me getting more tangled up in it all the time. I tell you, I was plumb scared. When the thing did stop for half a second I nearly broke up every wire and upright getting out of it.

—John T. Daniels, who snapped the famous photo of the Wright's first flight, describing what happened to the Wright flyer later that day.

There was something strange about the tall, gaunt figure. The face was remarkable, the head suggested that of a bird, and the features, dominated by a long, prominent nose that heightened the birdlike effect, were long and bony. . . . From behind the greyish blue depths of his eyes there seemed to shine something of the light of the sun. From the first moments of my conversation with him I judged Wilbur Wright to be a fanatic of flight, and I had no longer any doubt that he had accomplished all he claimed to have done. He seemed born to fly.

—The Daily Mail *newspaper, August 17, 1908.*

That Wilbur Wright is in possession of a power which controls the fate of nations is beyond dispute.

—*Major B. F. S. Baden-Powell, President of the Aeronautical Society of Great Britain.*

This morning at 3:15, Wilbur passed away, aged 45 years, 1 month and 14 days. A short life full of consequences, an unfailing intellect, imperturbable temper, great self-reliance and as great modesty, seeing the light clearly, pursuing it steadily, he lived and died.

—*Bishop Milton Wright, in his diary, May 30, 1912.*

Ignorance is the curse of God; knowledge is the wing wherewith we fly to heaven.

—*William Shakespeare.*

I know him well and he is just the kind of man to accomplish such an undertaking. He is apparently without fear and what he sets out to do he generally accomplishes. This recklessness makes him anything but a good aviator, however, for he lacks entirely the element of caution.

—*Wilbur Wright, speaking about Blériot after his first flight across the English Channel.*

A joke told repeatedly at aviation industry conferences puts a man and a dog in an airplane. The dog is there to bite the pilot if the man so much as tries to touch the controls; the pilot's one remaining job is to feed the dog. Many aviation veterans have heard the joke so many times that it is possible to tell those in the audience new to the industry by their laughter.

—*Gary Stix,* Scientific American, *July 1991.*

A plane is a bad place for an all-out sleep, but a good place to begin rest and recovery from the trip to the faraway places you've been, a decompression chamber between Here and There. Though a plane is not the ideal place really to think, reassess or reevaluate things, it is a great place to have the illusion of doing so, and often the illusion will suffice.

—*Shana Alexander.*

Flying has torn apart the relationship of space and time; it uses our old clock, but with new yardsticks.

—*Charles Lindbergh.*

Beware of men on airplanes. The minute a man reaches thirty thousand feet, he immediately becomes consumed by distasteful sexual fantasies which involve doing uncomfortable things in those tiny toilets. These men should not be encouraged, their sexual fantasies are sadly low-rent and unimaginative. Affect an aloof, cool demeanor as soon as any man tries to draw you out. Unless, of course, he's the pilot.

—*Cynthia Heimel.*

If I were reincarnated, I'd want to come back a buzzard. Nothing hates him or envies him or wants him or needs him. He is never bothered or in danger, and he can eat anything.

—*William Faulkner.*

Much talking is the cause of danger. Silence is the means of avoiding misfortune. The talkative parrot is shut up in a cage. Other birds, without speech, fly freely about.

—*Saskya Pandita.*

If I had to choose, I would rather have birds than airplanes.

—*Charles A. Lindbergh.*

I know of only one bird—the parrot—that talks; and it can't fly very high.

—*Wilbur Wright, declining to make a speech in 1908.*

Piloting

One can get a proper insight into the practice of flying only by actual flying experiments. The manner in which we have to meet the irregularities of the wind, when soaring in the air, can only be learned by being in the air itself.

—Otto Lilienthal, 1896.

What is chiefly needed is skill rather than machinery.

—Wilbur Wright, 1902.

Anyone can do the job when things are going right. In this business we play for keeps.

—Ernest K. Gann.

There are airmen and there are pilots: the first being part bird whose view from aloft is normal and comfortable, a creature whose brain and muscles frequently originate movements which suggest flight; and then there are pilots who regardless of their airborne time remain earth-loving bipeds forever. When these latter unfortunates, because of one urge or another, actually make an ascension, they neither anticipate nor relish the event and they drive their machines with the same graceless labor they inflict upon the family vehicle.

—Ernest K. Gann.

The way I see it, you can either work for a living or you can fly airplanes. Me, I'd rather fly.

—Len Morgan.

You'll be bothered from time to time by storms, fog, snow. When you are, think of those who went through it before you, and say to yourself, "What they could do, I can do."

—Antoine de Saint-Exupéry, Wind, Sand and Stars.

From knowing himself and knowing his airplane so well that he can come somewhere close to touching, in his own special and solitary way, that thing that is called perfection.

—Richard Bach, A Gift of Wings.

An airplane might disappoint any pilot but it'll never surprise a good one.

—Len Morgan.

Do not spin this aircraft. If the aircraft does enter a spin it will return to earth without further attention on the part of the aeronaut.

—First handbook issued with the Curtis-Wright flyer.

Rule books are paper—they will not cushion a sudden meeting of stone and metal.

—Ernest K. Gann, Fate is the Hunter.

The machine does not isolate man from the great problems of nature but plunges him more deeply into them.

—Antoine de Saint-Exupéry, Wind, Sand and Stars.

From a safety standpoint, in our view one of the things that we do in the basic design is the pilot always has the ultimate authority of control. There's no computer on the airplane that he cannot override or turn off if the ultimate comes. In terms of any of our features, we don't inhibit that totally. We make it difficult, but if something in the box should behave inappropriately, the pilot can say "This is wrong" and he can override it. That's a fundamental difference in philosophy that we have versus some of the competition.

—John Cashman, chief test pilot, Boeing 777.

I've never seen an airplane yet that can read the type ratings on your pilot's license.

—Chuck Boedecker.

Do not let yourself be forced into doing anything before you are ready.

—Wilbur Wright.

It is hard enough for anyone to map out a course of action and stick to it, particularly in the face of the desires of one's friends; but it is doubly hard for an aviator to stay on the ground waiting for just the right moment to go into the air.

—Glenn Curtiss, 1909.

In response to how he checked the weather, "I just whip out my blue card with a hole in it and read what it says: When color of card matches color of sky, FLY!"

—Gordon Baxter.

Instrument flying is an unnatural act probably punishable by God.

—*Gordon Baxter.*

The life of the modern jet pilots tends to be most unexpectedly lonely. . . . Foreign countries are places to reach accurately and to leave on time. Distance is a raw material to work with.

—*John Pearson,* Sunday Times, *February 4, 1962.*

I've flown every seat on this airplane; can someone tell me why the other two are always occupied by idiots?

—*Don Taylor.*

Son, never ask a man if he is a fighter pilot. If he is, he'll let you know. If he isn't, don't embarrass him.

—*The Great Santini,* Get Ready for a Fighter Pilot.

It is a good thing to learn caution from the misfortunes of others.

—*Publilius Syrus.*

He is most free from danger, who, even when safe, is on his guard.

—*Publilius Syrus.*

There's no such thing as a natural-born pilot.

—*General Charles "Chuck" Yeager.*

I have flown in just about everything, with all kinds of pilots in all parts of the world—British, French, Pakistani, Iranian, Japanese, Chinese—and there wasn't a dime's worth of difference between any of them except for one unchanging, certain fact: the best, most skillful pilot has the most experience.

—*General Charles "Chuck" Yeager.*

Most pilots learn, when they pin on their wings and go out and get in a fighter, especially, that one thing you don't do, you don't believe anything anybody tells you about an airplane.

—*General Charles "Chuck" Yeager.*

If you aren't sweating too much before a flight, you surely haven't asked enough questions. If you are not sweating just a little during the flight, you may not be attentive enough. And, if you are not sweating out the answers with all the experts you can think of after the flight, you may never find that very beautiful pearl in all that pig litter.

—*Corwin H. Meyer, Grumman test pilot, World War II.*

Flying is learning to throw yourself at the ground, and miss.

—*Douglas Adams,* The Hitchhikers Guide to the Galaxy.

When the weight of the paper equals the weight of the airplane, only then you can go flying.

—*Attributed to Donald Douglas (Mr. DC-n).*

Landing on the ship during the daytime is like sex; it's either good or it's great. Landing on the ship at night is like a trip to the dentist; you may get away with no pain, but you just don't feel comfortable.

—*Lieutenant Commander Thomas Quinn, USN.*

Get rid at the outset of the idea that the airplane is only an air-going sort of automobile. It isn't. It may sound like one and smell like one, and it may have been interior decorated to look like one; but the difference is—it goes on wings.

—*Wolfgang Langewiesche, first words of*
Stick and Rudder: An Explanation of the Art of Flying, *1944.*

The length of debate about a flight maneuver is always inversely proportional to the complexity of maneuver. Thus, if the flight maneuver is simple enough, debate approaches infinity.

—*Robert Livingston,* Flying the Aeronca.

A pilot must have a memory developed to absolute perfection. But there are two higher qualities which he also must have. He must have good and quick judgment and decision, and a cool, calm courage that no peril can shake.

—*Mark Twain, speaking about Mississippi River pilots.*

Keep the aeroplane in such an attitude that the air pressure is always directly in the pilot's face.

—*Horatio Barber, 1916.*

The pilot who teaches himself has a fool for a student.

—*Robert Livingston,* Flying the Aeronca.

The only time an aircraft has too much fuel on board is when it is on fire.

—*Sir Charles Kingsford Smith.*

The happily married man with a large family is the test pilot for me.

—*Nevil Shute,* Slide Rule.

Flexible is much too rigid; in aviation you have to be fluid.

—*Verne Jobst.*

If you can't afford to do something right, then be darn sure you can afford to do it wrong.

—*Charlie Nelson.*

Learning the secret of flight from a bird was a good deal like learning the secret of magic from a magician. After you know what to look for you see things that you did not notice when you did not know exactly what to look for.

—*Orville Wright.*

Anyone can hold the helm when the sea is calm.

—*Publilius Syrus.*

I hope you either take up parachute jumping or stay out of single motored airplanes at night.

—Charles A. Lindbergh to Wiley Post, 1931

Yeah, I knew a lot of those guys who parachute jumped at county fairs in the twenties and thirties. I just never knew any of them for very long.

—Fritz Orchard.

Never fly the "A" model of anything.

—Edward Thompson.

Never fly anything that doesn't have the paint worn off the rudder pedals.

—Harry Bill.

What kind of man would live where there is no daring? I don't believe in taking foolish chances, but nothing can be accomplished without taking any chance at all.

—Charles A. Lindbergh, at a news conference after his transatlantic flight.

Keep thy airspeed up, lest the earth come from below and smit thee.

—William Kershner.

Death is just nature's way of telling you to watch your airspeed.

—Anon.

Don't ever let an airplane take you someplace where your brain hasn't arrived at least a couple of minutes earlier.

—*Andy Anderson.*

When a prang seems inevitable, endeavour to strike the softest, cheapest object in the vicinity, as slowly and gently as possible.

—*Advice given to RAF pilots during World War II.*

The thing is, helicopters are different from planes. An airplane by its nature wants to fly, and if not interfered with too strongly by unusual events or by a deliberately incompetent pilot, it will fly. A helicopter does not want to fly. It is maintained in the air by a variety of forces and controls working in opposition to each other, and if there is any disturbance in this delicate balance the helicopter stops flying; immediately and disastrously. There is no such thing as a gliding helicopter.

This is why being a helicopter pilot is so different from being an airplane pilot, and why in generality, airplane pilots are open, clear-eyed, buoyant extroverts and helicopter pilots are brooding introspective anticipators of trouble. They know if something bad has not happened it is about to.

—*Harry Reasoner, 1971.*

Instrument flying is when your mind gets a grip on the fact that there is vision beyond sight.

—*U.S. Navy* Approach *magazine, circa World War II.*

Flying is done largely with the imagination.

—*Wolfgang Langewiesche,* Stick and Rudder: An Explanation of the Art of Flying, *1944.*

If you don't think you're the best pilot in the business, maybe you're in the wrong business. If you think you could never make a mistake, you are really in the wrong business.

—*Randy Sohn.*

Who was the best pilot I ever saw? You're lookin' at 'im.

—*Gordon Cooper, in the movie* The Right Stuff, *1983.*

There are two kinds of airplanes—those you fly and those that fly you . . . you must have a distinct understanding at the very start as to who is the boss.

—*Ernest K. Gann.*

And let's get one thing straight. There's a big difference between a pilot and an aviator. One is a technician; the other is an artist in love with flight.

—*Elrey Borge Jeppesen.*

I don't want monitors here. I want pilots. . . . Our whole philosophy is that the pilot is in charge of the airplane. We're very anti-automation here at this airline.

—*Greg Crum, system chief pilot, Southwest Airlines, 1996.*

Electronics were rascals, and they lay awake nights trying to find some way to screw you during the day. You could not reason with them. They had a brain and intestines, but no heart.

—*Ernest K. Gann,* The Black Watch, *1989.*

Nothing said I had to crash.

—*R. A. "Bob" Hoover, after hitting a telephone wire and losing 2 ft of wing in his P-51.*

In the Alaska bush I'd rather have a two hour bladder and three hours of gas than vice versa.

—*Kurt Wien.*

What is it in fact, this learning to fly? To be precise, it is "to learn NOT to fly wrong." To learn to become a pilot is to learn—not to let oneself fly too slowly. Not to let oneself turn without accelerating. Not to cross the controls. Not to do this, and not to do that. . . . To pilot is negation.

—*Henri Mignoet,* L'Aviation de L'Amateur; Le Sport de l'Air, *1934.*

The quality of the box matters little. Success depends upon the man who sits in it.

—*Baron Manfred Rittmeister von Richthofen, "The Red Baron."*

The successful pilot must have a quick eye and steady nerves.

—*W. J. Abbot.*

The airman must possess absolutely untroubled nerves.

—*Francis Collins.*

My first shock came when I touched the rudder. The thing tried to bite its own tail. The next surprise I got was when I landed; she stalled at a hundred and ten miles an hour.

—*Jimmy Haizlip, commenting on his only flight in the Gee Bee.*

Always keep an "out" in your hip pocket.

—*Bevo Howard.*

There's a lot of Hollywood bullshit about flying. I mean, look at the movies about test pilots or fighter pilots who face imminent death. The controls are jammed or something really important has fallen off the plane, and these guys are talking like magpies; their lives are flashing past their eyes, and they're flailing around in the cockpit. It just doesn't happen. You don't have time to talk. You're too damn busy trying to get out of the problem you're in to talk or ricochet around the cockpit. Or think about what happened the night after your senior prom.

—*Brigadier General Robin Olds, USAF.*

The Cub is the safest airplane in the world; it can just barely kill you.

—attributed to Max Stanley, Northrop test pilot.

I enjoyed my service flying very much. That is where I learned the discipline of flying. In order to have the freedom of flight you must have the discipline. Discipline prevents crashes.

—Captain John Cook, British Airways Concorde training captain.

I don't think I possess any skill that anyone else doesn't have. I've just had perhaps more of an opportunity, more of an exposure, and been fortunate to survive a lot of situations that many others weren't so lucky to make it. It's not how close can you get to the ground, but how precise can you fly the airplane. If you feel so careless with you life that you want to be the world's lowest flying aviator you might do it for a while. But there are a great many former friends of mine who are no longer with us simply because they cut their margins too close.

—R. A. "Bob" Hoover, television interview.

A pilot who doesn't have any fear probably isn't flying his plane to its maximum.

—Jon McBride, astronaut.

If you're faced with a forced landing, fly the thing as far into the crash as possible.

—R.A. Bob Hoover.

It occurred to me that if I did not handle the crash correctly, there would be no survivors.

—*Richard Leakey, after engine failure in a single-engine Cessna, Nairobi, Kenya, 1993.*

If an airplane is still in one piece, don't cheat on it. Ride the bastard down.

—*Ernest K.Gann, describing advice from "a very old pelican of an aviator,"* The Black Watch, *1989.*

Airshow flying is tough; it's even tougher if you do something stupid. Don't do nuthin dumb!

—*Ralph Royce.*

This thing we call luck is merely professionalism and attention to detail it's your awareness of everything that is going on around you it's how well you know and understand your airplane and your own limitations. Luck is the sum total of your abilities as an aviator. If you think your luck is running low, you'd better get busy and make some more. Work harder. Pay more attention. Study your NATOPS more. Do better preflights.

—*Stephen Coonts,* The Intruders.

The winds and the waves are always on the side of the ablest navigators.

—*Edward Gibbon.*

Map reading was not required. There were no maps. I got from place to place with the help of three things. One was the seat of my pants. If it left the plane, when the visibility was at a minimum, I was in trouble and could even be upside down. Another was the ability to recognize every town, river, railroad, farm, and, yes, outhouse along the route. The third? I had a few drops of homing pigeon in my veins

—Ken McGregor, early U.S. air mail pilot.

Any landing you can walk away from is a good one!

—Gerald R. Massie, USAAF photographer.
Written after the crash-landing of his B-17, 1944.

If you can fill out the yellow sheet with Jack Black in your hand instead of an I.V. in your arm, it was a good landing.

—Charlie Kisslejack, Commander, U.S. Navy, 1983.

A fierce and monkish art; a castigation of the flesh. You must cut out your imagination and not fly an airplane but regulate a half-dozen instruments. . . . At first, the conflicts between animal sense and engineering brain are irresistibly strong.

—Wolfgang Langewiesche, describing flying on instruments, A Flier's World, *1943.*

You've never been lost until you've been lost at Mach 3.

—Paul F. Crickmore, Lockheed SR-71: The Secret Missions Exposed, *1993.*

And he supposed it might not be the best of days. But then, he was flying the mails and was not expected to squat on the ground like a frightened canary every time there was a cloud in the sky. If a pilot showed an obvious preference for flying only in the best conditions he soon found himself looking for work. This was the way of his life and he had always ascended when others had found excuse to keep their feet on the ground.

—*Ernest K. Gann,* The Aviator.

Let all who build beware
The load, the shock, the pressure
Material can bear.
So, when the buckled girder
Lets down the grinding span,
The blame of loss, or murder,
Is laid upon the man.
Not on the Stuff—the Man!

—*Rudyard Kipling,* Hymn of Breaking Strain.

It's when things are going just right that you'd better be suspicious. There you are, fat as can be. The whole world is yours and you're the answer to the Wright brothers' prayers. You say to yourself, nothing can go wrong . . . all my trespasses are forgiven. Best you not believe it.

—*Ernest K. Gann, describing advice from "a very old pelican of an aviator,"* The Black Watch, *1989.*

Nobody who gets too damned relaxed builds up much flying time.

—*Ernest K. Gann, describing advice from "a very old pelican of an aviator,"* The Black Watch, *1989.*

I suddenly get a feeling—perhaps only a hint—of the ALONENESS of a 1930s transport pilot way up on the beak of this ancient pelican. This tiny cupola was not a "flight deck," all indirect lighting and softly chiming "systems," triply redundant captains murmuring their checklist incantations. This was one man stuck about as far out on the bowsprit of his ark as he could be without having his toes in the wind.

—*Stephan Wilkinson,* Flying *magazine, 50th anniversary issue, September 1977.*

The emergencies you train for almost never happen. It's the one you can't train for that kills you.

—*Ernest K. Gann, describing advice from "a very old pelican of an aviator,"* The Black Watch, *1989.*

If you want to grow old as a pilot, you've got to know when to push it, and when to back off.

—*General Charles "Chuck" Yeager.*

Great pilots are made not born. . . . A man may possess good eyesight, sensitive hands, and perfect coordination, but the end result is only fashioned by steady coaching, much practice, and experience.

—*Air Vice-Marshal James Edgar "Johnnie" Johnson, RAF.*

A pilot's business is with the wind, with the stars, with sand, with the sea. He strives to outwit the forces of nature. He stares with expectancy for the coming of dawn the way a gardener awaits the coming of spring. He looks forward to port as to a promised land, and truth for him is what lives in the stars.

—Antoine de Saint-Exupéry.

Accuracy means something to me. It's vital to my sense of values. I've learned not to trust people who are inaccurate. Every aviator knows that if mechanics are inaccurate, aircraft crash. If pilots are inaccurate, they get lost—sometimes killed. in my profession life itself depends on accuracy.

—Charles A. Lindbergh.

I learned that danger is relative, and that inexperience can be a magnifying glass.

—Charles A. Lindbergh.

Navigating by the compass in a sea of clouds over Spain is all very well, it is very dashing, but—you want to remember that below the sea of clouds lies eternity.

—Antoine de Saint-Exupéry, Wind, Sand and Stars.

There is no reason to fly through a thunderstorm in peacetime.

—Sign over squadron ops desk at Davis-Monthan AFB, Arizona, 1970.

There is no reason to fly through a thunderstorm.

—Sign over squadron ops desk at Ubon RTAFB, Thailand, 1970.

Conn McCarthy

Safety

Aviation in itself is not inherently dangerous. But to an even greater degree than the sea, it is terribly unforgiving of any carelessness, incapacity or neglect.

—Anon., dates back to a World War II advisory.

Beware, dear son of my heart, lest in thy new-found power thou seekest even the gates of Olympus. . . . These wings may bring thy freedom but may also come thy death.

—Daedalus to Icarus, after teaching his son to use his new wings of wax and feathers.

If you are looking for perfect safety, you will do well to sit on a fence and watch the birds; but if you really wish to learn, you must mount a machine and become acquainted with its tricks by actual trial.

—Wilbur Wright, 1901.

Insisting on perfect safety is for people who don't have the balls to live in the real world.

—Mary Shafer, NASA Ames Dryden.

The fundamental problem is government people—pointy-headed bureaucrats—telling people what to do. There is an environment in this city of people unwilling to admit their mistakes and move ahead. The attitude toward rule-making has been so curtailed that common sense recommendations now take years and years.

—Jim Hall, NTSB, 1996.

There are no new types of aircrashes—only people with short memories. Every accident has its own forerunners, and every one happens either because somebody did not know where to draw the vital dividing line between the unforeseen and the unforeseeable or because well-meaning people deemed the risk acceptable.

If politics is the art of the possible, and flying is the art of the seemingly impossible, then air safety must be the art of the economically viable. At a time of crowded skies and sharpening competition, it is a daunting task not to let the art of the acceptable deteriorate into the dodgers' art of what you can get away with.

—*Stephen Barlay,* The Final Call: Why Airline Disasters Continue to Happen, *March 1990.*

What is the cause of most aviation accidents?
Usually it is because someone does too much too soon, followed very quickly by too little too late.

—*Steve Wilson, NTSB investigator, Oshkosh, Wisconsin, August 1996.*

If the pilot survives the accident, you'll never find out what really happened.

—*Doug Jeanes.*

If there were no risks it probably would not be worth doing. I certainly believe an airplane is capable of killing you, and in that sense I respect it.

—*Steve Ishmael, NASA test pilot.*

Airplanes may kill you but they ain't likely to hurt you.

—Leroy Robert "Satchel" Paige, baseball player, circa 1959.

We tend to shy away from words like "dangerous," because we will not embark on anything unless it is truly thought out. But there is always an area of uncertainty . . . But we prefer to call it "high risk' rather than "dangerous."

—Squadron Leader Vic C. Lockwood, RAF,
principal fixed wing flying tutor, Empire Test Pilot School, 1985.

Flying is inherently dangerous. We like to gloss that over with clever rhetoric and comforting statistics, but these facts remain: gravity is constant and powerful, and speed kills. In combination, they are particularly destructive.

—Dan Manningham, Business and Commercial Aviation *magazine.*

Mix ignorance with arrogance at low altitude and the results are almost guaranteed to be spectacular.

—Bruce Landsberg, Executive Director of the AOPA Air Safety Foundation.

[Airplanes are] near perfect; all they lack is the ability to forgive.

—Richard Collins.

If the engine stops for any reason, you are due to tumble, and that's all there is to it!

—Clyde Cessna, founder of Cessna aircraft.

A 10 cent fuse will protect itself by destroying the $2,000 radio to which it is attached.

—Robert Livingston, Flying the Aeronca.

No matter how interested individual employees might be, or what assistance a manufacturer offers, or how insistent a certificating authority might be—none of these factors will have a significant effect on safety without support from top management.

—John O'Brian, ALPA's Engineering and Air Safety Department.

The cost of solving the Comet mystery must be reckoned neither in money nor in manpower.

—Sir Winston Churchill.

The airlines spell safety with a dollar sign and the FAA practices regulation by death.

—Patricia Robertson Miller, Chicago Sun-Times, *August 1, 1979.*

The explosion of the "Challenger," after twenty-four consecutive successful shuttle flights, grounded all manned space missions by the U.S. for more than two years. The delay barely evoked comment . . . But contrast the early history of aviation, when 31 of the first 40 pilots hired by the Post Office died in crashes within six years, with no suspension of service.

—C. Owen Paepke.

I know how to never have another Challenger. I know how to never have another leak, and never to screw up another mirror, and that is to stop and build some shopping centers in the desert.

—J. R. Thompson, NASA deputy administrator.

If we die, we want people to accept it. We are in a risky business, and we hope that if anything happens to us it will not delay the program. The conquest of space is worth the risk of life.

—Astronaut Virgil I. "Gus" Grissom. On January 27, 1967, astronauts Grissom, White, and Chaffee died from a flash fire aboard an Apollo spacecraft.

I'd hate to see an epitaph on a fighter pilot's tombstone that says, "I told you I needed training." . . . How do you train for the most dangerous game in the world by being as safe as possible? When you don't let a guy train because it's dangerous, you're saying, "Go fight those lions with your bare hands in that arena, because we can't teach you to learn how to use a spear. If we do, you might cut your finger while you're learning." And that's just about the same as murder.

—Colonel "Boots" Boothby, USAF.

I am a history major. I believe that the past is prologue. The archives bear that out. Most major aircraft accidents are not acts of God. In our recommendations we try to take what we have learned and correct situations so it shouldn't happen again.

—Jim Hall, NTSB, 1996.

Whenever we talk about a pilot who has been killed in a flying accident, we should all keep one thing in mind. He called upon the sum of all his knowledge and made a judgment. He believed in it so strongly that he knowingly bet his life on it. That his judgment was faulty is a tragedy, not stupidity. Every instructor, supervisor, and contemporary who ever spoke to him had an opportunity to influence his judgment, so a little bit of all of us goes with every pilot we lose.

—Anon.

Take nothing for granted; do not jump to conclusions; follow every possible clue to the extent of usefulness. . . . Apply the principle that there is no limit to the amount of effort justified to prevent the recurrence of one aircraft accident or the loss of one life.

—U.S. Air Force, Accident Investigation Manual.

Where the water meets the sky, it was just fire.

—Jarreau Israel, eyewitness to the crash of
TWA Flight 800 into the waters off New York's Long Island, 1996.

Its important not to focus so much on the statistics, but [on people's] perceptions.

—Federico Peña, U.S. Transportation Secretary, quoted in USA Today *December 22, 1994.*

I have flown ValuJet. ValuJet is a safe airline, as is our entire aviation system.

—Federico Peña, May 12, 1996.

I go out of my way to stay off commuter planes. I have skipped conferences because I would not fly on marginal airlines (and because of many mishaps, I also avoided flying on ValuJet).

—*Mary Schiavo, U. S. DOT Inspector General,* Newsweek, *May 20, 1996.*

I strongly take exception to her comments. When we say an airline is safe to fly, it is safe to fly. There are no gray areas.

—*David R. Hinson, Federal Aviation Administrator, under oath to a Senate committee.*

Yes, the airline is safe. I would fly on it. It meets our standards.

—*David R. Hinson, Federal Aviation Administrator, May 12, 1996.*

ValuJet was later grounded for not meeting FAA standards.

Every accident, no matter how minor, is a failure of the organization.

—*Jerome Lederer.*

The alleviation of human error, whether design or intrinsically human, continues to be the most important problem facing aerospace safety.

—*Jerome Lederer.*

The desire for safety stands against every great and noble enterprise.

—*Cornelius Tactitus, circa* A. D. *100.*

Every year, more people are killed by injuries caused by donkeys than those caused by aircraft.

—found in The Toastmaster, *official journal of Toastmasters International.*

Of the major incentives to improve safety, by far the most compelling is that of economics. The moral incentive, which is most evident following an accident, is more intense but is relatively short lived.

—Jerome Lederer.

Corporate culture has a very real influence on the attitudes and performance of the people within an organization. There is no question in my mind that management decisions and actions, or more frequently, indecisions and inactions, cause accidents.

—John Lauber, NTSB.

The high level of safety achieved in scheduled airline operations lately should not obscure the fact that most of the accidents that occurred could have been prevented. This suggests that in many instances, the safety measures already in place may have been inadequate, circumvented or ignored.

—International Civil Aviation Organization, Accident Prevention Manual, *1984.*

Complacency or a false sense of security should not be allowed to develop as a result of long periods without an accident or serious incident. An organization with a good safety record is not necessarily a safe organization.

—*International Civil Aviation Organization,* Accident Prevention Manual, *1984.*

The ATR-72 airplane must be shown to be capable of continued safe flight and landing when operating in any weather conditions for which operation is approved, including icing conditions. . . . As the FAA cannot issue a new or amended type certificate for an airplane with a known unsafe design feature, ATR must provide data to show that the [icing] problems experienced with the ATR-42 will not be present on the ATR-72.

—*Federal Aviation Administration, 1989.*

"Ice" was a four-letter word to ATR pilots.

—*Stephan A. Fredrick,* Unheeded Warning, *1996.*

If I had to choose in which type of aircraft not to set foot, I would probably be more worried by the amazing number of Boeing crashes during the last few years, which were generally caused by technical failures, or remain unexplained . . .

—*Jean-François Bosc, Ecole Nationale de l'Aviation Civile, in reply to questions about the safety of the French-built ATR-72, Toulouse, France, October 21, 1996.*

Why does one want to walk wings? Why force one's body from a plane to make a parachute jump? Why should man want to fly at all? People often ask these questions. But what civilization was not founded on adventure, and how long could one exist without it? Some answer the attainment of knowledge. Some say wealth, or power, is sufficient cause. I believe the risks I take are justified by the sheer love of the life I lead.

—Charles Lindbergh.

When it comes to testing new aircraft or determining maximum performance, pilots like to talk about "pushing the envelope." They're talking about a two dimensional model: the bottom is zero altitude, the ground; the left is zero speed; the top is max altitude; and the right, maximum velocity, of course. So, the pilots are pushing that upper-right-hand corner of the envelope. What everybody tries not to dwell on is that that's where the postage gets canceled, too.

—Admiral Rick Hunter, USN.

For they had learned that true safety was to be found in long previous training, and not in eloquent exhortations uttered when they were going into action.

—Thucydides, The History of the Peloponnesian War, *circa 404* B.C.

A ship in harbor is safe—but that is not what ships are for.

—John A. Shedd.

NASA

Space

That's one small step for [a] man; one giant leap for mankind.

—Neil Armstrong, the first words spoken by a man walking on another heavenly body,
July 20, 1969. He didn't realize he had messed up the line until after he got to Earth,
according to the book Chariots for Apollo *by Charles R. Pellegrino and Joshua Stoff*
(not the NASA technical memorandum on the same subject and with an identical title).
When presented with a plaque by the builders of the lander he pointed out their mistake
in failing to include the "a," and was told that the word was not in the tapes.
He insisted (at that time) that he had said it.

Whoopee! Man, that may have been a small one for Neil, but it's a long one for me.

—Pete Conrad, the shortest Apollo astronaut, upon becoming the third man to walk on the moon.

A little levity is appropriate in a dangerous trade.

—Walter M. Schirra, Jr.

Contact light. Okay, engine stop.

—Edwin "Buzz" Aldrin, Jr., the first words after landing on the moon.

Houston, Tranquillity Base here, The Eagle has landed.

—Neil Armstrong, 3:18 P.M. Houston time, July 20, 1969.

Roger, Tranquillity. We copy you on the ground. You've got a bunch of guys about to turn blue. We're breathing again. Thanks a lot.

—Charlie Duke, Houston Capcom, replying.

Man looks aloft, and with erected eyes
Beholds his hereditary skies.

—Ovid.

Ad astra, per aspera.

"To the stars, through hardship."
Motto of the Royal Flying Corps, formed on April 13, 1912.

I believe that this nation should commit itself to achieving the goal, before this decade is out, of landing a man on the moon and returning him safely to earth. No single space project in this period will be more impressive to mankind or more important for the long-range exploration of space. And none will be so difficult or expensive to accomplish.

—President John F. Kennedy, May 25, 1961.

We go into space because whatever mankind must undertake, free men must fully share.

—President John F. Kennedy, 1962.

. . . the United States was not built by those who waited and rested and wished to look behind them. This country was conquered by those who moved forward, and so will space.

—President John F. Kennedy, 1962.

But why, some say, the moon? Why choose this as our goal? And they may well ask, why climb the highest mountain? Why, 35 years ago, fly the Atlantic? Why does Rice play Texas?

—President John F. Kennedy, 1962.

If I could get one message to you it would be this: the future of this country and the welfare of the free world depends upon our success in space. There is no room in this country for any but a fully cooperative, urgently motivated all-out effort toward space leadership. No one person, no one company, no one government agency has a monopoly on the competence, the missions, or the requirements for the space program.

—President Lyndon B. Johnson.

The United States this week will commit its national pride, eight years of work and $24 billion of its fortune to showing the world it can still fulfill a dream. It will send three young men on a human adventure of mythological proportions with the whole of the civilized world invited to watch—for better or worse.

—Rudy Abramson, Los Angeles Times, *July 13, 1969.*

It is the policy of the United States that activities in space should be devoted to peaceful purposes for the benefit of all mankind.

—U.S. Space Act of 1958.

Don't tell me that man doesn't belong out there. Man belongs wherever he wants to go—and he'll do plenty well when he gets there.

—*Dr. Wernher von Braun.*

It's human nature to stretch, to go, to see, to understand. Exploration is not a choice, really; it's an imperative.

—*Michael Collins.*

To go places and do things that have never been done before—that's what living is all about.

—*Michael Collins.*

Space, the final frontier. These are the voyages of the Starship Enterprise. Its five-year mission: to explore strange new worlds, to seek out new life and new civilizations, to boldly go where no man has gone before.

—*Captain James T. Kirk, beginning of every episode of the original TV series* Star Trek.

I had the ambition to not only go farther than man had gone before, but to go as far as it was possible to go.

—*Captain Cook*

What good is the Moon? You can't buy it or sell it.

—*Ivan F. Boesky, Wall Street broker convicted of insider trading.*

A-OK full go.

—*Commander Alan Shepard, Jr., on blast-off of rocket carrying him aloft as America's first man in space, May 5, 1961. Defined as an engineering term for "double OK" or perfect, it became a U.S. idiom for "everything is going smoothly" and was later attributed by the Associated Press (*The New York Times, *July 31, 1963) to Lieutenant Colonel John Powers, public spokesman for astronauts.*

This is the greatest week in the history of the world since the Creation.

—*President Richard M. Nixon, to the* Apollo 11 *crew, July 24, 1969.*

It's a strange, eerie sensation to fly a lunar landing trajectory—not difficult, but somewhat complex and unforgiving.

—*Neil Armstrong.*

Now I want to partially close the hatch, making sure not to lock it on my way out.

—*Edwin "Buzz" Aldrin, Jr., leaving the lunar lander.*

Here Men From Planet Earth First Set Foot Upon The Moon July 1969 A.D. We Came In Peace For All Mankind.

—*Plaque left on the moon.*

It's different, but it's very pretty out here. I suppose they are going to make a big deal of all this.

—*Neil Armstrong, transmitting from the moon.*

Neil and Buzz, I am talking to you by telephone from the Oval Office at the White House, and this certainly has to be the most historic telephone call ever made . . . Because of what you have done, the heavens have become a part of man's world. As you talk to us from the Sea of Tranquillity, it inspires us to redouble our efforts to bring peace and tranquillity to Earth.

—President Richard M. Nixon.

It is a monster, that rocket. It is not a dead animal; it has a life of its own.

—Guenter Wendt.

This beast is best felt. Shake, rattle, and roll. We are thrown left and right against our straps in spasmodic little jerks. It is steering like crazy, like a nervous lady driving a wide car down a narrow alley, and I just hope it knows where it's going, because for the first ten seconds we are perilously close to that umbilical tower.

—Michael Collins.

And now 'tis man who dares assault the sky . . .
And as we come to claim our promised place,
Aim only to repay the good you gave,
And warm with human love the chill of space.

—Professor Thomas G. Bergin, Yale University, Space Prober. *This was the first poem to be launched into orbit about the Earth. It was inscribed on the instrument panel of a satellite called* Traac *launched from Cape Kennedy on November 15, 1961.*

The mass gross absence of sound in space is more than just silence.

—*Eugene Cernan.*

You almost wish you could turn off the COMM and just appreciate the deafening quiet.

—*Russell Schweickart.*

What the space program needs is more English majors.

—*Michael Collins.*

It's not quite as exhilarating a feeling as orbiting the earth, but it's close. In addition, it has an exotic, bizarre quality due entirely to the nature of the surface below. The earth from orbit is a delight—offering visual variety and an emotional feeling of belonging "down there." Not so with this withered, sun-seared peach pit out of my window. There is no comfort to it; it is too stark and barren; its invitation is monotonous and meant for geologists only.

—*Michael Collins,* Carrying the Fire.

To set foot on the soil of the asteroids, to lift by hand a rock from the Moon, to observe Mars from a distance of several tens of kilometers, to land on its satellite or even on its surface, what can be more fantastic? From the moment of using rocket devices a new great era will begin in astronomy: the epoch of the more intensive study of the firmament.

—*Konstantin E. Tsiolkovsky, "father of Russian astronautics," 1896.*

The greatest gain from space travel consists in the extension of our knowledge. In a hundred years this newly won knowledge will pay huge and unexpected dividends.

—Dr. Wernher von Braun.

The real friends of the space voyager are the stars.

—Jim Lovell, Apollo 13 *commander.*

"In the Beginning god created the Heaven and the Earth. And the Earth was without form and void. And darkness was upon the face of the Deep. . . . And God saw that it was Good. . . . " And from the crew of Apollo 8, we close with good night and a Merry Christmas. And God bless all of you, all of you on the good Earth.

—Flight crew of Apollo 8 *flight crew, Christmas Eve, 1968.*

The view of the moon that we've been having recently is really spectacular. It fills about three-quarters of the hatch window, and of course we can see the entire circumference even though part of it is in complete shadow and part of it is in earthshine. It's a view worth the price of the trip.

—Neil Armstrong.

This blowing dust became increasingly thicker. It was very much like landing in a fast-moving ground fog.

—Neil Armstrong.

Any sufficiently advanced technology is indistinguishable from magic.

—*Sir Arthur C. Clarke*, Profiles of the Future, *1962*.

Program Alarm, it's a 1202.

—*Neil Armstrong, during first lunar descent.*

Roger, we're GO on that alarm

—*Charlie Duke, Houston Capcom, replying to Neil. The computer overflowed several more times.*

Armstrong sitting in the commander's seat, spacesuit on, helmet on, plugged into electrical and environmental umbilicals, is a . . . machine himself.

—*Norman Mailer.*

Hey Houston, we've had a problem here.

—*Jack Swigert,* Apollo 13 *command module pilot.*

Say again please.

—*Jack Lousma, Houston Capcom, replying to Jack.*

Err Houston, we've had a problem. We've had a main B bus undervolt.

—*Jim Lovell, Apollo 13 commander.*

I don't believe any pair of people had been more removed physically from the rest of the world than we were.

—Edwin "Buzz" Aldrin, Jr.

Outer space is no place for a person of breeding.

—Lady Violet Bonham Carter.

We believe that when men reach beyond this planet, they should leave their national differences behind them.

—President John F. Kennedy, news conference, February 21, 1962.

... space is for everybody. It's not just for a few people in science or math, or for a select group of astronauts. That's our new frontier out there, and it's everybody's business to know about space.

—Christa McAuliffe, December 6, 1985.

One test result is worth one thousand expert opinions.

—Dr. Wernher von Braun.

A human being is the best computer available to place in a spacecraft . . . It is also the only one that can be mass produced with unskilled labor.

—Dr. Wernher von Braun.

Our two greatest problems are gravity and paper work. We can lick gravity, but sometimes the paper work is over whelming.

—Dr. Wernher von Braun.

Man is the animal that intends to shoot himself out into interplanetary space, after having given up on the problem of an efficient way to get himself five miles or so to work and back each day.

—Bill Vaughan, quoted in Reader's Digest, *January 1956.*

Ten years ago the moon was an inspiration to poets and an opportunity for lovers. Ten years from now it will be just another airport.

—Emmanuel G. Mesthene.

Earth is too small a basket for mankind to keep all its eggs in.

—Robert A. Heinlein.

Once you get to earth orbit, you're halfway to anywhere in the solar system.

—Robert A. Heinlein.

I have argued flying saucers with lots of people. I was interested in this: they keep arguing that it is possible. And that's true. It is possible. They do not appreciate that the problem is not to demonstrate whether it's possible or not but whether it's going on or not.

—Dr. Richard Feynman.

Sometimes I think we're alone in the universe, and sometimes I think we're not. In either case the idea is quite staggering.

—Sir Arthur C. Clarke.

Houston, Apollo 11 . . . I've got the world in my window.

—Michael Collins.

I really didn't appreciate the first planet [earth] until I saw the second one. . . . I cannot recall the [moon's] tortured surface without thinking of the infinite variety the delightful planet earth offers.

—Michael Collins.

We used to joke about canned men, putting people in a can and seeing how far you can send them and bring them back. That's not the purpose of this program . . . Space is a laboratory, and we go into it to work and learn the new.

—John Glenn, Jr.

To get your name well enough known that you can run for a public office, some people do it by being great lawyers or philanthropists or business people or work their way up the political ladder. I happened to become known from a different route.

—John Glenn, Jr.

It's a very sobering feeling to be up in space and realize that one's safety factor was determined by the lowest bidder on a government contract.

—Alan Shepard, Jr.

You're in charge but don't touch the controls.

—Shannon Lucid, recounting what the two Russian cosmonauts
told her every time they left the Mir space station for a spacewalk, 1996.

The only thing it would be nice to have more of would be M & M's.

—Shannon Lucid, after 6 months on the Mir Space Station, 1996.

From now on we'll live in a world where man has walked on the moon. It's not a miracle, we just decided to go.

—Jim Lovell.

Space isn't remote at all. It's only an hour's drive away, if your car could go straight upwards.

—Sir Fred Hoyle.

If we did not have such a thing as an airplane today, we would probably create something the size of NASA to make one.

—H. Ross Perot.

Why our space program? Why, indeed, did we trouble to look past the next mountain? Our prime obligation to ourselves is to make the unknown known. We are on a journey to keep an appointment with whatever we are.

—*Gene Roddenberry.*

Those who came before us made certain that this country rode the first waves of the industrial revolution, the first waves of modern invention and the first wave of nuclear power. And this generation does not intend to founder in the backwash of the coming age of space. We mean to be part of it—we mean to lead it.

—*President John F. Kennedy.*

For forty-nine months between 1968 and 1972 two dozen Americans had the great good fortune to briefly visit the Moon. Half of us became the first emissaries from Earth to tread its dusty surface. We who did so were privileged to represent the hopes and dreams of all humanity. For mankind it was a giant leap for a species that evolved from the stone age to create sophisticated rockets and spacecraft that made a Moon landing possible. For one crowning moment, we were creatures of the cosmic ocean, an epoch that a thousand years hence may be seen as the signature of our century.

—*Edwin "Buzz" Aldrin, Jr.*

Conn McCarthy

Women Fly

So many men now have lost their lives in airplane accidents that individual addition [sic] to the long list of their names have ceased to cause any really deep emotions except in the minds of their relatives and friends. When a woman is the victim however the feeling of pity and horror is as strong as was that produced by the first of these disasters to men and though there is at present no expectation that aviation should be abandoned by men because of the recognized dangers, the death of Miss Bromwell is almost sure to raise in many minds at least the question if it would not be well to exclude women from a field of activity in which there [sic] presence certainly is unnecessary from any point of view.

—The New York Times, *editorial, 1921.*

Had I been a man I might have explored the Poles or climbed Mount Everest, but as it was my spirit found outlet in the air. . . .

—*Amy Johnson, essay in* Myself When Young, *1938.*

So I accept these awards on behalf of the cake bakers and all of those other women who can do some things quite as important, if not more important, than flying, as well as in the name of women flying today.

—*Amelia Earhart.*

Please know I am quite aware of the hazards. I want to do it because I want to do it. Women must try to do things as men have tried. When they fail their failure must be but a challenge to others.

—*Amelia Earhart, in her last letter to her husband, 1937.*

Women will never be as successful in aviation as men. They have not the right kind of nerve.

—*Maurice Hewlet, the first English lady to solo an airplane.*

Women must pay for everything. . . . They do get more glory than men for comparable feats, But, also, women get more notoriety when they crash.

—*Amelia Earhart.*

I didn't know a lot about Amelia before I started [flying]. And as a woman and a pilot, I should have known more.

—*Linda Finch, prior to starting out on a flight retracing Amelia Earhart's last journey, 1997.*

Flying is a man's job and its worries are a man's worries.

—*Antoine de Saint-Exupéry,* Wind, Sand and Stars.

It is now possible for a flight attendant to get a pilot pregnant.

—*Richard J. Ferris, President of United Airlines.*

She's decisive, she's aggressive, she's proven she's capable with high-performance jets. We look for people with the capability to think on their feet and to be able to lead a team of people. We look for the best pilots out there, and if they happen to be women, great, but we're just looking for the best.

—*David Leestma, director of flight crew operations,*
Johnson Space Center, with regard to astronaut Susan Still, 1997.

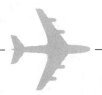

Last Words

The most frequent last words I have heard on cockpit voice-recorder tapes are, "Oh Shit,' said with about that much emotion. There's no panic, no scream, it's a sort of resignation: we've done everything we can, I can't think of anything else to do and this is it.

—Frank McDermott, partner in McDermott Associates, specialists in cockpit voice recorders.

Sacrifices must be made.

—Otto Lilienthal, one of the main sources of inspiration for the Wright Brothers.
He died August 10, 1896, from injuries sustained 2 days earlier in a crash of
one of his hang gliders. Original German: "Opfer müssen gebracht werden."

Yes I will succeed and I'll make some money, unless I break my neck.

—Eugène Lefebure, just before his fatal crash in a Wright Flyer,
quoted in Le Petit Parisien, *September 9, 1909.*

Higher, ever higher.

—Georges Chavez, after crashing his Blériot airplane on his trailblazing flight
over the Alps, September 1910. His words became the motto of the Peruvian Air Force.

What's the hurry? Are you afraid I won't come back?

—Baron Manfred Rittmeister von Richthofen, "The Red Baron," last recorded words,
in reply to a request for an autograph as he was climbing into the cockpit of his plane.

I have a feeling that there is just about one more good flight left in my system and I hope this trip is it. Anyway when I have finished this job, I mean to give up long-distance "stunt" flying.

—Amelia Earhart, departing from Los Angeles for Florida on May 21, 1937. It was the start of her last flight.

Did he not clear the runway—that Pan American?

—Flight engineer William Schreuder, KLM, March 27, 1977, just prior to the worst aviation crash ever, the collision of two B-747s on the ground in the Canary Islands.

Hey—what's happening here?

—Captain Robert Loft, Eastern Airlines Flight 401, December 29, 1972, last recorded words before crashing into the Florida Everglades.

Ma, I love yah.

—Last recorded words from PSA 182, after a fatal midair crash with a Cessna over San Diego, September 25, 1978. The unidentified voice was one of the pilots, the flight engineer, or a company pilot riding the jumpseat.

God, look at that thing!
That don't seem right, does it?
That's not right.

—First Officer Roger Pettit, during takeoff roll, Air Florida Flight 90, January 13, 1982.

Larry! We're going down, Larry!
I know it.

—First Officer Roger Pettit and Captain Larry Wheaton, last words recorded on
Air Florida Flight 90, close to the 14th Street Bridge, Washington, D.C., January 13, 1982.

Go with throttle-up . . . uh-oh . . .

—Francis R. Scobee, Commander of the Space Shuttle Challenger, last recorded words.

I've got a problem [uttered at 3000 ft. while in an inverted flat spin].
I've really got a problem [at 1500 ft.].

—Art Scholl, last recorded words before fatal crash
while filming a stunt sequence for the movie Top Gun.

Do you hear the rain? Do you hear the rain?

—Jessica Dubroff, 7-year-old "pilot" speaking to her mother by telephone as the engines
revved for takeoff. She and her flight instructor crashed minutes later in rough weather, 1996.
The Federal Aviation Regulations were later changed to stop such record flights by small children.

*I*ndex

ABOUT THE AUTHOR

Dave English really is English. He grew up just outside of London, England, and studied physics at the University of Warwick. After moving to the States, he fulfilled a lifetime dream by learning to fly. He currently works as an airline pilot based out of Chicago O'Hare, while living in Milwaukee, Wisconsin. An editor at *Airways* and *IFR* magazines, he has also written articles for several other aviation publications.